ECOVILLAGES AROUND THE WORLD

20 Regenerative Designs for Sustainable Communities

ECOVILLAGES
AROUND THE WORLD

FINDHORN PRESS

Findhorn Press

One Park Street

Rochester, Vermont 05767

findhornpress.com

Findhorn Press is a division of Inner Traditions International

A CIP record for this title is available from the Library of Congress

ISBN 978-1-84409-743-2 (print)
ISBN 978-1-84409-763-0 (e-book)

Text design and layout by Kurt Rodahl Hoppe

Printed and bound in India by Replika Press Pvt. Ltd.
This book was typeset in Sofia Pro and Charter.
Picture credits: see pages 174-175.

To send correspondence to the one of the authors of this book, mail a first-class letter to the author c/o Inner Traditions • Bear & Company, One Park Street, Rochester, VT 05767, and we will forward the communication.

Contents

Tribute to Hildur

Hildur was my partner in life and in work for over 50 years before her untimely passing in 2015. It is difficult to sum up in a few words the legacy of such a unique person. She had first and foremost a loving heart—an unconditional love for everything living—humans, animals, nature in all its many facets. Many have suggested that she was the very embodiment of the Mother Earth concept. In fact, she once experienced being the whole Earth during a holotropic breathwork session with Stanislav Grof.

But she was not a passive observer. Rather she was a grass roots and social justice activist. You name it, she was there; peace movement, anti-nuclear movement, women's lib movement, environmental movement, ecovillage movement; always on the front lines. But she was much more—a very intelligent woman with a degree in Law and postgraduate studies in Cultural Sociology; a very spiritual woman, being a leading member of the global organization Sahaj Marg; an energetic extrovert, always full of fun with a wall-to-wall smile and a new initiative underway. It was hard to keep up with her.

Most of all for me personally, she was a loving wife and the best mother one could wish for one's children.

I was so fortunate to know her. And I miss her very much.

Ross Jackson
Chairman, Gaia Trust

Tribute to Hildur Jackson

Hildur Jackson was a visionary, mentor and sister for many. For us she was a pathfinder of our times, not only because of her rich personal life journey, but for the effects her vision has had on the lives of countless others. From the brilliance of her engagement in connecting regenerative community experiments and weaving the Global Ecovillage Network, to her bravery in stating the need for a 'new' education and engendering Gaia Education, the ripple effects of her life are immeasurable.

She believed that we can take the future in our hands and, through the power of community, consciously restore our social and natural environments. She called on us all to integrate our spiritual with our scientific and practical aspects to grow into a full expression of who we truly are. Her friendship with Wangari Maathai was an expression of her ongoing deep love for Africa—and it was this love that engendered her strong support for the emergence of GEN Africa as a new region within GEN in 2012.

Over the last decade, in her role of Gaia Education Publications Coordinator Hildur launched the 4 Keys for Sustainable Communities everywhere on the Planet—social design, ecological design, economic design and worldview. To complete the series she wanted the 5th Key to be a beautiful coffee table book about the 20 best practice designs for ecovillages, as a legacy of her passion for communities. She compiled a list of both veteran ecovillage projects and new innovative ones and shared it with Ross, her beloved husband and lifelong companion, and us before she died in the autumn of 2015.

This book is a visual tribute to her unique womanhood, compassionate heart, inspired thinking and unexpected acts of beauty and inclusion through which she rendered service to our world.

We are walking in Hildur's footsteps, she was our mentor and sister—and it is an honour to continue the work she initiated.

May East, CEO, Gaia Education
Kosha Joubert, Executive Director,
Global Ecovillage Network

Ecovillage Voices

"Whereas the dominant themes in world politics remain the arms trade, industrialization, development, and growth, there is a subculture made up of a growing and bonded network of like-minded fellow travellers who are trying to short-circuit the industrial structure, to live closer to the land, to consume less, to use less nonrenewable energy, and to exchange their wares and skills through trade, cooperatives and mutual aid, and an exchange of information that is not based on profit.
There are groups in such far-flung places as Ethiopia, Guatemala, India, and Swaziland; in every country in Europe; in New Zealand and Australia. Unlike the dominant world politics in which control is channeled through several enormously powerful nerve centers, the alternative is evolving a dynamic in which the periphery becomes the centre, and the centre is everywhere."

John and Nancy Jack Todd—New Alchemists and builders of bioshelters

"This, indeed, is the greatest gift of the ecovillage movement: the delinking of levels of consumption and well-being. Their most subversive message is that beyond a certain standard of living (that almost all households in Western Europe have long since passed), greater well-being is to be had not through the consumption of more stuff, but by way of sharing and the building of meaningful relationships within a human-scale community."

Martin Stengel, Sieben Linden

"Ecovillages and/or intentionally created sustainable communities are human-scale full-featured settlements in which human activities are harmlessly integrated into the natural world in a way that is supportive of healthy human development, and can be successfully continued into the indefinite future."

Robert and Diane Gilman

"When we walk the path of cooperation with nature, we will one day recognize that the word 'paradise' is no longer only a religious term but a life-task."

Leila Dregger, Tamera

"Our systems thinking approach aims to illustrate that the design of cities, towns or villages in this manner is the best way to achieve environmental sustainability. One-planet living can be achieved by designing each place to support—or even provide an abundance of—basic needs for a known population. This approach also has the potential to alter the economics of housing as future residents will not simply be purchasing 'a home' or 'a property' in the sense that these concepts are currently understood. Residents of future ecovillages will be stewards of an ecosystem. They will therefore be 'buying into' a set of responsibilities to collectively manage the systems that provide them with their basic needs."

Ilan Arnon, Tasman Ecovillage

"I see communal living as a default setting i.e., it's the most natural way for human beings to cohabitate. It should be the norm, and of course it was, up until the Industrial Revolution some 300 years ago. For millennia beforehand, we mostly lived as fully interdependent, mutually supportive members of tribes, hamlets, villages and towns. And we lived sustainably! If present-day communal living has a purpose at all, then perhaps it's to remind us of this now forgotten fact."

Graham Meltzer, Findhorn

"Beginning with a quick look at the etymology of the word, community. The word is derived, in part, from the Latin, communitas, meaning 'fellowship'. So community is, by definition, about the bonds and ties between members of a given communal group. It is about their relationships. This is the nub of community life whether it be within an intentional community (e.g., ecovillage, commune, kibbutz, monastery, cohousing, etc.) or in society at large. Additional etymological roots come from French *communité*, meaning 'commonness', and again, from Latin *communis*, meaning 'shared by all or many'. So holding in common or sharing, whether it be of land and infrastructure, or values and agreements, is also fundamental to community. These two aspects, relationships and sharing, are essentially what define community."

Graham Meltzer, Findhorn

"Ecovillages are the newest and most potent kind of intentional community. They unite two profound truths: human life is at its best in small, supportive, healthy communities, and the only sustainable path for humanity is in the recovery and refinement of traditional community life."

**Dr Robert J Rosenthal,
Professor of Philosophy,
Hanover College**

"A model for the future needs not only new technology and a healthy ecology, but also people who are able to use these tools in a meaningful way. It needs people who have learned how to stay together even during conflicts, solving them in non-violent and creative ways and remaining committed to solidarity even in difficult times. Community knowledge is the foundation of social sustainability."

Leila Dregger, Tamera

Ecovillage Voices

"I first worked as a Rural Development advisor in Tanzania in 1986. At that time, we were all searching for ways to transform agriculture—to create wealth for rural communities, to ensure food security, to provide secure long-term livelihoods for their families, and to above all give pride, respect and dignity to the poorest of the poor. This was our search for the Holy Grail. Twenty-five years on we think we have found it. Within an incredible three years, Chololo is becoming a household name for innovation and success in the world of rural development. One of the most fragile and vulnerable rural communities in Tanzania is showing the way. I feel personal pride and satisfaction in being associated with this story. The story is not yet over, but the inspirational achievements speak for themselves. I salute all those who continue to make this happen. I salute in particular all the Chololo villagers who have taken risks, changed their practices and become true Ambassadors for rural development in Africa. Long may they flourish."

Tim Clarke, Former EU Ambassador

"While there is no one definition of an 'ecovillage', these villages are characterized by striving to take a systemic approach to integrating the human environment with the natural environment. Thus ecovillages aim to develop green buildings, grow organic food, use renewable energy, create a strong sense of community, use a participatory governance system, and teach what they are learning through practical, hands-on methods."

Liz Walker, Ithaca

Declaration of Global Interdependence and Sustainable Settlements

- All Life is one and intimately interrelated. We influence each other, and are dependent on each other, all over the planet. We are one with nature, plant and animal systems. Every single person is responsible for the whole and can influence it. This is a new spiritual paradigm, which replaces a materialistic one, with great implications. Formulated about 100 years ago by science in the West, but ages ago in the East by perennial philosophy, it is now ready to become the foundation of a new global culture.
- Consciousness is foremost, and we are as such manifestations of global consciousness.
- We all share the possibility of access to global consciousness and co-creation of it. Our light may be dimmed by impressions from this and earlier lifetimes, and we may need to clean away these old impressions and outdated thoughts for our light to shine, and for us to gain access to global consciousness.
- The purpose of life on Earth is the continuation of 3 billion years of evolution. It is the fundamental purpose behind all political, economic and social structures.
- Humans can live aligned in support of this purpose and create societies and communities in harmony with this purpose, and with spiritual and natural laws. Global justice, listening to nature and respecting her, are fundamental.

- In doing so, settlements, villages and communities will be holographic reflections of the whole—'as above so below'.
- Groups creating such communities will create settlements that are sustainable—spiritually, ecologically, economically and socially. A new economic system, new laws and new technologies will be invented and adapted to support this endeavour.
- The role of planners and architects will be to facilitate the vision and needs of such groups. Planners will need an overview of all the knowledge necessary to build such communities. Gaia Education and the four-week holistic introduction, Ecovillage Design Education (EDE) provides this overview.
- Architecture, renewable energy supply, gardening, local food production, businesses, common facilities, social structure, local economy will be one integrated totality planned together as 'ecovillages'.
- Ecovillages are possible in all local communities all over the world. They will happen once the materialistic worldview of a global market is abandoned voluntarily, because it cannot solve the problems, or because we are forced by nature or other circumstances.
- An ecovillage lifestyle is also a peace initiative, as it will not destroy nature or species, create pollution, regional wars or struggles over land, resources (oil) or power.

Hildur Jackson

Why Ecovillages?
by Frederica Miller, editor

Welcome on an amazing journey! This book will take you around the world, and through time, to 20 wonderful ecovillage projects. Each project is unique and different, but united in its commitment to providing solutions to the global social and environmental challenges that confront us. Ecovillagers meet the climate goals we all have to meet, and they are doing it with innovation, enthusiasm and joy!

Voluntary simplicity, luxurious simplicity, doing more with less, collaborative consumption, sharing or gift economy, these words describe the creation of a lifestyle that makes it easy to consume less, live more lightly on the planet, and give space for all living creatures.

There are initiatives taken by grass-root groups, individuals, top-down projects, projects that have changed lifeless deserts into green oases. There are those that provide income and food for the climate vulnerable, that bring water back to drought-ridden areas, that question the fundamental flaws of our economic system by creating local currencies, that provide the affluent middle class with a lower carbon footprint. It seems there are ecovillages all over the world that have solutions for any and all environmental, social, economic and cultural challenges our era faces. It is possible, and they are doing it! Not only are they providing practical solutions, they are doing it in a non-violent way, with a profound respect for our individuality and diversity, while promoting communal solutions.

Social sustainability is one of the keys to transformation. In an era that is constantly searching for the quick techno fix, ecovillages are going a different route. Without a socially sound and biological grounding, technology alone cannot solve our era's enormous challenges. People are the problem, and the problem is the solution. If we don't fundamentally change our lifestyles, we will undoubtedly continue to destroy our and many other living beings' habitats. To change we need to DO things differently—action leads to transformation—we need practical laboratories where we can create a fundamentally different culture. Ecovillages are certainly some of the most intensive living laboratories we have!

Many of the foremost ecovillages have been scrutinized by academic research to see if they actually deliver what they promote—in practice. Available research shows that all ecovillages dramatically reduce their ecological footprint (a way of measuring our need for resources). The most successful have reached the goals we all have to reach—to live well within a finite planet. In the affluent Western world, this means reducing the ecological footprint by anything from a third to a tenth of today's norm. For the exploited parts of the world, it means increasing their footprint without following the destructive example set by the conventional Western world.

This book provides a view into the future. Its many diverse examples show solutions in particular times and places. In all their experimental imperfection, ecovillages are perfect examples of what we need more of! I have chosen to place them chronologically, with the oldest projects first and the newest last. This shows an interesting development, from projects that originally had a more spiritual origin, like Findhorn, or socially responsible origins like Solheimar, to fully-fledged ecologically planned villages that provide ready-built homes for anyone and everyone interested in making a difference, like Hurdal and Permatopia. To me this shows that the concept of ecovillages is no longer a marginal movement, but a powerful force for change. A change that is happening not by revolution but by gentle evolution—the natural way.

"If your goal is to enrich the society that you are a part of, and to have your own needs met by the society you have created, then the principle that works is: the less I take for myself, the more I will get."

Ruben Khachatryan, Camphill Village, Norway.

The United Nations 17 Sustainable Development Goals introduced in 2015 are almost a checklist of ecovillage practices: people—planet—prosperity—partnership and peace. In this sense, we are all in need of development. In fact, the richest and most affluent need it more than any others do!

I want to thank all of those who so generously have contributed to this book. As individuals, I have come to appreciate each and every one of you! As representatives of the amazing projects you represent and have presented you are, as you have been careful to tell me, only one of many voices. Depending on whom you ask, very different stories can be told. This to my mind is another confirmation of the richness inherent in all ecovillages.

To me your existence is a huge inspiration; you are demonstrating how we can create supportive, tolerant human communities, which also respect our mother earth and all its diversity. You are a beacon of light and hope in the challenging times we inhabit!

The choice of projects in this book is entirely Hildur's. Her extensive and first-hand knowledge through many years has ensured an impressive choice of ecovillages. Thank you Hildur and Ross for giving me the opportunity to edit this book.

Using a term like "the best" is always difficult, not least in this context. There is only one criterion for "the best" ecovillage and that is whether you actually answer the challenges you face where you are. In this sense, these ecovillages are only a small number of examples, which represent the many thousands of wonderful initiatives constantly popping up all over the world. In a modern day context, all these initiatives represent the BEST of what we humans can do! Thank you. May diversity reign and the seeds of this book multiply!

Solheimar Iceland

The cultivation of man and nature

by Gudmundur Armann

Sólheimar is a unique community with a particular emphasis on the cultivation of man and nature. The community was founded in 1930 by Sesselja H. Sigmundsdóttur (1902–1974) as a children's home, but has since evolved into a community where the needs of people, with or without disabilities, and others who need support, are provided for in a creative environment with artistic, cultural and environmental dimensions.

In 1930, Sólheimar village was the first place in the Nordic countries to practise biodynamic/organic agriculture, and the first to care for people with disabilities outside institutions. The theories of Rudolf Steiner had a formative influence on the community. Parts of the community still look to Steiner's philosophy and those who shape today's society regarding environmental and anthroposophical theories, but Sólheimar has also evolved on its own terms.

Edda in the greenhouse.

The village's main strengths lie in the many possibilities that each individual has for employment, and the social dynamic within the community, where all residents, people with or without disabilities, volunteers and others are involved on equal terms.

Sólheimar has had a great emphasis on being open and welcoming guests, while simultaneously making sure that residents can enjoy their privacy at home. About 35,000 people visit Sólheimar annually to learn about the environment, enjoy refreshments from certified organic ingredients, attend art and cultural events, buy organically certified vegetables grown on site, or bread and pastries from the bakery.

Aerial view of Sólheimar.

Edda and Loa packing tomatoes.

The ideology of Sólheimar is 'reverse integration', where the community is built up based on the needs of individuals with disabilities and other minorities, and where the able-bodied adapt to their needs. Environmentalism is an important and influential part of daily life in the community and shapes its development. Sólheimar is a global community which receives both short-term and long-term volunteers. Sólheimar cooperates with American educational organizations and university students can take a one-semester course, Sustainability Through Community, while residing on site.

Sustainability and personal development are interwoven into the daily lives of residents and constantly evolving. Although many aspects of Sólheimar are very successful, it is not a perfect society and there are many things regarding environmental issues that can be improved.

The advantage of Sólheimar is that many things are really well done, and we are systematically and continuously developing our community and trying to do better.

Welcome to Sólheimar!

Location: Southern Iceland, ca. 80 km (50 mi) from the capital city of Reykjavik.
Established: 1930.
Area: 264 ha (652 ac), including 10 ha (25 ac) of gardens and 190 ha (470 ac) of forest.
Population: 120 people. Average age is 40 years old. Male/female: 50/50.
Housing: 52 homes, with a total living space of 4504 m^2 (48,480 ft^2), mostly single houses with a home for the elderly and two communal houses.
Common facilities: A total of ca. 10,000 m^2 (108,000 $ft^{2)}$ museum, kitchen, bakery, canteen and office space, sports hall, gym, physiotherapy room, café, shop, church, storage building, guesthouses, eco centre, art workshops and greenhouses.

Exhibition in Sólheimar.

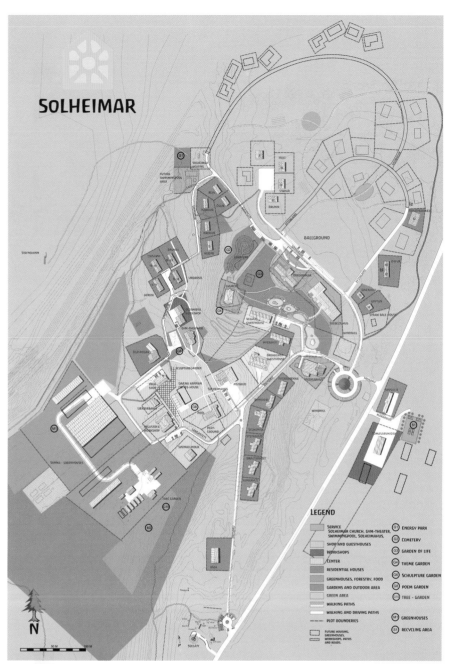

From top, morning circle, Sólheimar choir and troll.

SOLHEIMAR

LEGEND

SERVICE
SOLHEIMAR CHURCH, GYM-THEATER,
SWIMMINGPOOL, SOLHEIMAHUS,

SHOP AND GUESTHOUSES

WORKSHOPS

CENTER

RESIDENTIAL HOUSES

GREENHOUSES, FORESTRY, FOOD

GARDENS AND OUTDOOR AREA

GREEN AREA

WALKING PATHS

WALKING AND DRIVING PATHS

PLOT BOUNDERIES

FUTURE HOUSING,
GREENHOUSES,
WORKSHOPS, PATHS
AND ROADS.

G1 ENERGY PARK

G2 CEMETERY

G3 GARDEN OF LIFE

G4 THEME GARDEN

G5 SCHULPTURE GARDEN

G8 POEM GARDEN

G10 TREE - GARDEN

M1 GREENHOUSES

S1 RECYCLING AREA

N

Below, the permaculture garden at Sólheimar.

Solheimar **23**

Findhorn Scotland
Love in action

by Graham Meltzer

Findhorn ecovillage is one of the most comprehensively developed anywhere, which is why we've been called 'the mother of all ecovillages'. We have evolved a holistic, integrated community model incorporating many ecological, social, cultural, economic and spiritual elements, such as: numerous shared buildings including a Community Centre that serves meals twice a day; a performing arts centre and visual arts centre; eco-housing of many different types (for example, attached, detached, mobile, straw, recycled); extensive gardens and a large food growing area; our own wind farm that produces roughly the amount of electricity we use; an on-site biological sewage treatment system; our own sustainably harvested woodland; a centralized woodchip boiler that distributes heat to more than a dozen community buildings; a carpool, shop and much more. We also have land and buildings elsewhere that render our range of resources even more comprehensive: Cluny Hill, five miles away (a 100-room late Victorian building that was once a hotel and spa) and a retreat house on the mystical west coast Isle of Iona. On the island of Erraid, next to Iona, we have a satellite community of a dozen or so members as well as workshop and guest facilities. A fleet of

shuttle buses transports members and guests between these locations.

The Findhorn Foundation and Community, as we prefer to be known, began unintentionally in November 1962 when the three founders, Eileen Caddy, Peter Caddy and Dorothy Maclean first settled in Findhorn with the Caddys' three children, Christopher, Jonathan and David. It could perhaps be said that a mini-community began ten years earlier when the three came together to deepen their shared spiritual journey, which was later to become the foundation stone of the community. In Peter's words, "During the previous ten years every action of our lives had been directed by guidance from the voice of God within." God spoke to them via messages that Eileen channelled in meditation, and so it was, after life harshly dumped them in the Findhorn Bay Caravan Park, that Eileen's guidance continued to shape every aspect and determine every detail of their lives. Because they were flat broke, surviving only on Peter's unemployment benefit of eight pounds per week, they started a garden in which to grow food. Soon after, Dorothy also began to receive messages, which she attributed to the plant kingdom. Through 'inner listening', she first contacted what she referred to as the deva or spirit of the garden

From top: Strawbale house, the only one in The Park; highly crafted eco-houses built from local stone; Traigh Bhan, retreat house on the Isle of Iona.

Location: Northeast Scotland
Established: 1962
Area: 180 ha (445 ac), 80 ha (198 ac) nature reserve (coastal dune system), 16 ha (40 ac) forest, 4 ha (10 ac) agriculture, 20 ha (50 ac) housing and village.
Population: About 250 permanent residents with up to 300 additional visitors for workshops, conferences and events.
Housing: 60 detached houses, 30 row houses, 10 apartments, 30 residential caravans and 10 ecomobiles. Included are two cohousing projects of 25 and 6 households.
Common facilities: Community Centre, Visitor Centre, performance and visual arts centres, ecological housing, wind farm, biological sewage system, biomass boilers, gardens, food production, sustainably managed woodland, carshare, B&Bs, craft studios, offices and business premises, workshop and teaching rooms, meditation sanctuaries, whole food and craft shop.

pea, then went on to communicate with devas of many more plant species as well as elementals and unseen beings of different kinds. Most of these messages were practical—where, when and what to plant, how to make compost, and much more. Peter would enact the guidance, applying 'work is love in action' to develop the garden, all the while seeking to 'co-create with the intelligence of nature'. Soon enough they began to enjoy remarkable success, growing an abundance of oversized, healthy organic produce in extremely unlikely conditions—barren soil in a very hostile environment. With the aid of modest publicity that Peter disseminated, visitors started arriving to see and experience what was going on. Some of them stayed on and, as they say, the rest is history. Even though the founders never intended or even imagined founding a community, one formed around them...and it grew, and it grew. From then until the present day, Findhorn community members have continued

Wind farm comprising four wind generators totalling 750 kW.

to experience extraordinary connection and interaction with unseen elements of the natural world and what is called 'the subtle realms'. These days, we are both a spiritual community premised on the three basic tenets established by our founders (i.e., inner listening, work as love in action, and co-creation with nature) and an ecovillage. We began as just the former, and adopted the latter identity in the 1980s. I think it's fair to say that we have a kind of dual personality—we see ourselves and are viewed by others as either or both a spiritual community and an ecovillage. For the most part, these two aspects of our culture co-exist in harmony. It's been said that they are two sides of the one coin. As our identity has matured through the 1990s and into the new millennium we've applied more and more effort to living sustainably, replacing the ageing caravans with energy efficient and healthy buildings, producing more of our own renewable energy and substantially reducing our carbon footprint.

Today, some 700 members live in and around The Park, as the main ecovillage campus is called. The community is diverse in its demography, complex in its organization and rich in its social and cultural milieu. For various reasons, we have never had a single master plan for the development of housing, community buildings and infrastructure, although there have been numerous less formal development plans involving a plethora of architects, planners and member consultation processes. We inherited a caravan park full of unhealthy, unsustainable buildings—caravans being very poorly insulated and ventilated. Many of them still exist, well past their use-by date. Historically, the village has evolved in an ad hoc manner, according to 'God's will' some would say, but also as a result of the everchanging flow of people and resources that have come through here over time. Design and development decisions have often been made in the same way that most of our other decisions are made i.e., by 'attunement', which is a facet of our spiritual practice here. This too, contributes to the non-linearity of our development process.

An attunement is a mini-meditation of sorts. We 'tune in' to our inner wisdom in meditation and are guided by what emerges. The concept was developed at Findhorn in the 1970s by David Spangler, sometimes referred to as the fourth founder of our community. Attunement, he says, requires a 'repatterning of one's inner state so as to align or connect with spirit'. It involves shifting consciousness to

Nature Sanctuary, a meditation space
built from found and salvaged materials.

"Why do we need time at the sanctuary?
… It is a place where we can come together collectively to consciously generate the energies of love, light, peace, joy, wisdom and divine power, which we do in silence. Then at the end these energies can be sent out, not only to those around us or to the community alone, but to the world. This is where we become 'world servers' and link up with the 'network of light'."

Eileen Caddy

allow greater sensitivity and openness to subtle phenomena. In Findhorn we utilize attunement many times a day, sitting with colleagues in silence for a few minutes before beginning a work shift or meeting, for example. I'm very much reminded in these moments of the core purpose of our community established by Peter, Eileen and Dorothy some 50 years ago, which is to build a 'transformational field'. 'Going within' is probably the most quintessential aspect of the culture here in the Findhorn Foundation and Community.

Changing the world one heart at a time

We generally welcome anyone and everyone into the community, no matter what their background or belief system. This is both our greatest advantage (because it brings variety) and our biggest challenge (it can cause disharmony). But in my opinion, it's what makes being in our community such a joy on a minute-to-minute, day-to-day basis. It delivers a much-cherished richness to our social and cultural life.

In economic terms, the community comprises numerous different organizations: charities, non-profits, for profits and social enterprises of all kinds—from freelance artists to solar panel manufacturers, building companies to the whole foods and craft shop. Findhorn has its own local currency (the EKO) that many of our community businesses accept in place of the pound sterling. The original and largest organization is the Findhorn Foundation, an educational charity that holds most of the workshops, conferences and events and owns most of the community land and buildings. The Foundation runs programmes continuously throughout the year, hosting some 4,000 residential guests annually, many of whom undergo profound life-changing experiences. We are, as we like to say, 'changing the world one heart at a time'.

Up until 2017 the Foundation was an income sharing or egalitarian subset of the community as a whole, i.e. its 100+ staff received exactly the same financial remuneration irrespective of their role

Cluny Hill, workshop centre and accommodation for members and guests.

or contribution. We are now exploring paying people with more responsibility and long-term commitment a slightly higher rate. Outside the Foundation, community members make their living in a variety of ways. Some are employed, many are teachers, therapists and craftspersons and others are retired. In this, and most else, we are very diverse.

The Foundation and the community have together developed over many years what we call the Common Ground. The Findhorn Foundation website states that the Common Ground is 'a living document, a code of conduct, and used as a tool for transformation for ourselves, the community and the world'. It lists 14 agreements that represent values to which we all hold: spiritual practice, service, personal growth, personal integrity, respecting others, direct communication, reflection, responsibility, non-violence, perspective, co-operation, peacekeeping, agreements and commitment. One way or another, everyone joining the community will at some point sign up to these agreements.

The 'magic of Findhorn'

There are three main aspects that I most appreciate about my life in Findhorn: the people, the place and the culture. By far the most important is the people, or more specifically, my relationships with them, fashioned as they are by the Common Ground. The place and the culture, to my mind, provide the context for those relationships. The integration of people, place and culture can result in a deeply embodied experience of what I can only describe as a 'field of love'. I often feel as if I'm immersed in a culture where love is freely, constantly and generously expressed. Our core spiritual aspirations are open-heartedness and consciousness. If we humans interrelate thoughtfully and with an open heart, then magic happens—defences are dropped, aggression melts away and space opens for compassion, empathy and love to flow. This is the 'magic of Findhorn' as far as I'm concerned.

Work is love in action

I believe that shared meals are the single most important 'ritual' in the daily life

Right, Living Machine, a biological sewage treatment plant, and Cullerne Gardens, food production in polytunnels.

of almost all intentional communities. Certainly, at Findhorn, our community meals (available twice daily) are central to the culture and a critical component of the social glue. Of course there is something powerfully symbolic about sharing a meal, both with members of one's 'tribe' and with guests. I am no anthropologist but I would guess that 'breaking bread' holds this value (and has forever done so) for almost every cultural group, anywhere in the world. Our kitchen staff dream up and produce fabulous meals in our Community Centre. The food is vegetarian and the ingredients are, as much as we can make them, fresh, organic, local and seasonal. There are always dairy-free, gluten-free and sugar-free alternatives. It fills me with pride in my community to be reminded at every meal just how much trouble we take to cater for diversity, to meet the needs of every individual, and in this way, demonstrate inclusivity and caring for each other. Our community meals are prepared with love—the kitchen crews demonstrate our key ethos, 'work is love in action,' every single shift. In Findhorn, we believe that human values such as love and connection, and spiritual values such as the unity of everything, are more important than things material, particularly

the accumulation of material things. In contemporary society, there is a burgeoning phenomenon called the 'sharing economy' or 'collaborative consumption'. One value of sharing is people connecting—it builds social capital; it brings people together; it makes people happier. A sustainable society is also one in which we choose positive behaviours that make us feel happier, more connected and more disposed to help others. At Findhorn we already do a lot of this kind of sharing. We collectively own land, numerous community buildings and facilities. Many community members, myself included, are able to live in smaller dwellings because we share communal facilities such as laundry, guest rooms, office and workshop space.

From top: Universal Hall, recycled whisky barrel house, 50th birthday gathering.

When I need a car for a short journey, I can choose from the vehicles in our community carpool. The carpool has about 100 community members (and a few non-community members) who currently share 14 vehicles including three that are fully electric.

Our cultural life is a key ingredient of the community glue here, along with our spirituality and ecological concerns and practices. These three lifestyle strains are separate and distinct, but also blend together beautifully to help build community and strengthen relationships. Cultural life in Findhorn can be as full and rich as one wishes it to be. Most creative activities occur on campus. We're a community that loves

to dance and we enjoy regular sessions, classes and workshops in many dance forms, for example 5 Rhythms, contact improvisation, sacred dance, traditional Scottish dancing, Biodanza, trance dance and disco. All of these forms are celebrations of life, love and the joy of being human. The purpose is to enjoy dancing together in a non-competitive way, to learn that it is possible for everyone to dance together, to feel self-confident in a group that is supportive rather than critical, and to feel in contact with the Earth, spirit and each other through the different qualities of each dance. It is also used as a tool to channel healing energy for the dancers and for the rest of the planet.

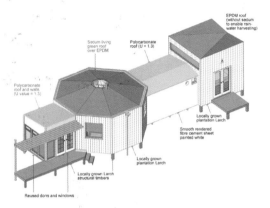

The ecomobile home

Ten years ago, I had the privilege of designing and building my home. It's what we call an 'ecomobile,' i.e., a residential building built to the regulations governing caravans and mobile homes, but built in situ with a high ecological specification. The house is a vehicle for sustainable living. Designed for a single person or a couple, it offers high levels of comfort and amenity whilst enabling the occupants to minimize their environmental footprint. The building is about half the size (per person) of the average UK dwelling. Small dwellings require fewer materials to construct, less energy to heat, and can hold less 'stuff'. Beyond material considerations, however, it offers a supportive setting for 'voluntary simplicity'—a less consumerist, more conscious and environmentally benign lifestyle characterized by ease and beauty.

The building is located in an area of high ecological value and sensitivity where full-grown specimen trees form a nature corridor linking two areas of wildlife habitat. Because the building has a small footprint and 'touches the ground lightly' (resting as it does on just a few pad footings), it can be set amongst the trees with minimal impact. The house is designed as a space of retreat, a place of psychological and spiritual nurture. For most of us who work in the Findhorn Foundation, community life is very busy, often intense. Every day we interact closely with guests, many of whom we meet as strangers. This can be very demanding and of course also very rewarding, but for me, it requires a place to return to in the evening where I can recharge my batteries. My home provides the setting for a contemplative life; a place where body, mind and soul may find peace.

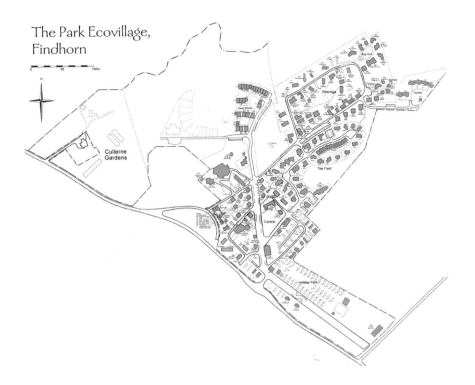

The Park Ecovillage,
Findhorn

Auroville India
A centre for human unity

Text and photos by MARTI

Auroville is an intentional community that was founded as a centre for human unity 50 years ago. It was inspired by the French spiritual visionary, Mira Alfassa, known as the Mother, and by the teachings of the great Indian sage, Sri Aurobindo. The Charter of Auroville reads, "Auroville belongs to no one in particular, but to humanity as a whole." Auroville is a cluster of communities that are evolving into an eco-town. Auroville's size and high degree of cultural diversity make the community unique within the GEN network.

Auroville receives about 1000 visitors a day. Many of them come to see the Matrimandir, a place for concentration and peace that defines the centre of the township. The Matrimandir has a large banyan tree as a companion and is surrounded by carefully designed gardens. Visitors also come to participate in workshops and training programmes in Auroville's many environmental, educational and commercial units.

Auroville excels in many different fields, including architecture and design, reforestation, experimental education, renewable energy, food security, water harvesting, appropriate technology, sustainable energy, small businesses and handicrafts, comparative philosophy, village outreach, art and culture. In Auroville all life is considered yoga, and Auroville is seen as a living laboratory for evolution and transformation of consciousness. The Mother called Auroville 'the city the Earth needs'.

Auroville Art Centre.

The Auroville Charter

Auroville belongs to nobody in particular.
Auroville belongs to humanity as a whole.
But, to live in Auroville, one must be a
willing servitor of the divine consciousness.

Auroville will be the place of an unending
education, of constant progress, and a youth
that never ages.

Auroville wants to be the bridge between the
past and the future. Taking advantage of all
discoveries from without and within,
Auroville will boldly spring towards
future realizations.

Auroville will be a site of material and
spiritual researches for a living embodiment
of an actual human unity.

Location: Tamil Nadu, South India, the Bay of Bengal.
Established: 1968
Area: 24 600 000 ha (26km radius) 61 000 000 ac (16 mi radius) in a dry forest region and green belt area.
Population: More than 2500 people from 60 nations, 90 different linguistic groups.
Common facilities: Visitors Centre, Industrial Zone, Matrimandir and Garden of Peace, International Zone, Cultural Zone with schools, youth centre, amphitheatre, concert hall, Centre Field with solar kitchen, and solar café.

Left, Galaxy Vision.

Above: Tibetan Centre courtyard. Right: Bharat Nivas Auditorium.
Below: Overview of the city plan.

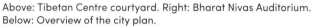

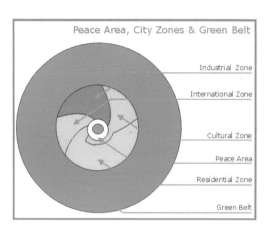

The city the earth needs

The Auroville Township has a master plan, inspired largely by the Mother and by Auroville's chief architect Roger Anger, who was a student of the French architect Le Corbusier. The township is divided into four zones where each serves a specific purpose: residential, cultural, industrial, and international. The Matrimandir and the gardens of peace are in the middle. The design of the community resembles a spiral or galaxy that can be seen from space. Lacing the zones is a curving greenbelt with forests, parks and green corridors.

The community's architectural and landscape design is dictated by location,

planning and circumstances. The city of Pondicherry, with more than one million inhabitants, is only eight kms away, so there is considerable pressure from India's growing population and from real estate developers, as the land has never been completely consolidated. The International Zone is home to regional pavilions. Each pavilion is dedicated to the cultural specificities of its region, with countries in Europe, Africa, Asia, and the Americas. The Bharat Nivas area, or Indian pavilion, has a large auditorium, which seats thousands and is dedicated to Sri Aurobindo. There is also a Tibetan Pavilion and library that have been blessed by His Holiness the Dalai Lama. On New Year's Eve, the community gathers in

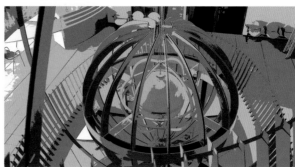

Auroville Town Hall.

candlelight under Tibetan prayer flags to pray and meditate for peace for the coming year. The Tibetan Centre houses exhibitions, workshops, meetings and cultural exchanges with Tibetan trainees and visitors.

Auroville's Visitors Centre is also in the International Zone and has a restaurant, outdoor amphitheatre and boutiques with beautiful handicrafts that are all locally produced in Auroville. From time to time, Aurovillians prepare community cultural dinners there, featuring food from countries as diverse as Spain and Korea.

The Industrial Zone is home to many of Auroville's small industries. There are more than 150 small businesses in Auroville ranging from organic food,

local handicrafts and clothing to solar, wind, ferro cement technology and dynamized water.

One of the clothing design units, Upasana, has outreach projects aimed at providing work for the threatened silk-weavers in Benares. Upasana also provides a demand for natural cotton in India. Another unit, Well Paper, is organized to help village women become skilled craftspeople and owners of their own units, and to help them with the marketing of the products they make out of recycled newspapers.

The Auroville Earth Institute, a group of dedicated architects and builders, does experimental work in compressed earth architecture and has exported its technology throughout Southeast

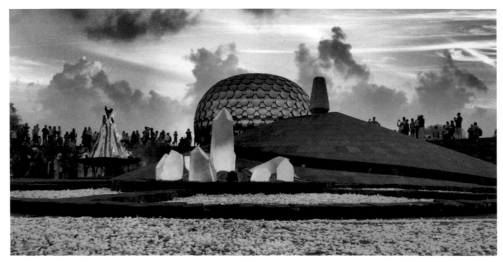

Matrimandir house of contemplation.

Citadine Living Wall.

Asia where homes, mosques, temples, and institutions have been built using compressed earth blocks. Village Action, a group dedicated to improving conditions in the surrounding villages, works with Aurovillian and local educators to provide better housing conditions, clean water, and medical and dental aid to Auroville's immediate neighbours.

Auroville's Town Hall is located on the edge of the Cultural Zone and features a movie hall, meeting rooms and an exhibition space for town planning. Nearby is a collective living experiment called Citadines for long-term Aurovillians and a youth community for young people who come from abroad to do apprenticeship programmes in Auroville. The Cultural Zone houses most of Auroville's schools. Auroville has about a dozen schools ranging from village-affiliated schools, to experimental schools, to a high school that prepares for the international baccalaureate and boasts Harvard and Cambridge University graduates. Formal and informal education is always based on hands-on experiential learning,

Above: Maison des Jeunes youth volunteer capsule. Top right: Auroville Solar Kitchen. Bottom right: Indian Pavilion.

sports and active collective projects. Auroville youth have a deep passion for the environment and are getting ready to take over leadership positions from Auroville's ageing pioneer population.

The Cultural Zone also has a youth centre, a music studio, an outdoor amphitheatre and a concert hall. Cultural programmes reflect the social and ethnic diversity of the community and vary from traditional Tamil music and theatre to modern jazz and Shakespearean plays and improvisational theatre.

The solar kitchen and solar café near Centre Field are where many community members meet for lunch and dinner. The solar kitchen has an immense solar bowl on its roof where the sun's heat is channelled into steam ovens in the kitchen below. The bowl is the largest of its kind and can prepare food for about a

thousand people. A bulletin board at the entry to the cafeteria features a long list of workshops, concerts, films and events in the community at any given time.

The Residential Zone has many types of architecture, with homes made out of materials such as wood, bamboo, earth blocks and ferro cement. Some of the neighbourhood communities have waste water systems integrated into their building and landscaping plans. There are many stand-alone homes, apartments of varying design and houses integrated into the environment.

The Greenbelt—a thick band of forest and tree growth—covers about one third of Auroville. For nearly 50 years, the Forest Group has planted and maintained it with an emphasis on mixed forests and indigenous species. More than 3 million trees have been planted since Auroville

Sadana Forest reforestation settlement with the community meeting hall and playground.

started and these trees slowly began to come to life on a dry, eroded stretch of coastline, which was clear-cut more than a century ago by colonial powers.

Auroville is in a tropical dry forest region and its foresters provide expertise, seeds, and seedlings to other communities in India, Nepal and parts of South East Asia. Green corridors have been established, where wildlife and people can pass on foot, bicycle and on horseback. The numbers of birds and animals have increased significantly in the area in recent years.

About half the land that Auroville needs to realize its dream, however, still remains to be purchased and the price of real estate and village encroachments have made it difficult to consolidate all the land that was in the original plan.

Living in Auroville is not without its challenges. The climate is hot and humid most of the year. Roads are becoming increasingly congested. The community is continually reinventing itself in terms of internal governance. However, the beauty of Auroville is that most Aurovillians are committed to living ecologically, realising their spiritual aspirations, finding climate change solutions and working on a collective transformation of consciousness. Sri Aurobindo talked about the emergence of a new species as a result of a decisive quantum shift in awareness. This may be coming soon as we face new ways of healing Mother Earth and responding to the global challenges of planetary survival.

Damanhur Italy
Deep contact with nature

by Macaco Tamerice

Damanhur is a Federation of spiritual Communities, nestled in the alpine foothills of northern Italy, in Piedmont, mainly in a beautiful green valley called Valchiusella. Damanhur is multilingual and very active, with its own constitution, culture, art, music, currency, school, environmental stewardship and use of science and technology. Many guests come to visit every year and Damanhurians are happy to share their experiences and research in many fields.

Damanhur's philosophy is based on positive thinking, action and the idea that every individual desires to contribute to the evolution of humanity, through inner awareness and transformation. Our planet 'Gaia' is understood to be a sensitive being, to whom Damanhurians feel deeply connected, and environmental stewardship is a natural consequence.

Damanhur has slowly developed into a United Nations award-winning sustainable eco-community and is actively implementing many of the UN's Sustainable Development Goals. The 600 resident citizens live in 25 different 'Nucleo Communities'. They use the community as an opportunity for self-refinement, through exchange with others, positive thinking, embracing diversity and change, and pursuing common and personal dreams

with a sense of humour and adventure. Everyday life is where philosophy comes alive in art, culture, family life, labour and politics and researching the subtle energies of the universe. In order to express the connection to nature, but also to emphasize change and humour, many Damanhurians choose to adopt an animal and plant name. So when you come and visit you might talk to *Gazza Solidago* (Magpie Solidago), *Ornitorinco Platano* (Platypus Plane Tree) or *Condor Girasole* (Condor Sunflower).

In Damanhur the contact with the forces of nature, guided by the desire to re-establish a harmonious relationship with life on our beautiful planet, has always been an active field of research. Damanhurians feel that humans are part of a spiritual ecosystem with forces and intelligences. It is important to establish a conscious contact with these, as it is with all life forms present in the environmental ecosystem that surrounds us. Our human evolution is inextricably linked to the reunification with the physical and subtle forces that inhabit our planet. Plants and nature spirits are inhabitants of our world, and a large part of the research in Damanhur is dedicated to experiencing a deep contact with them. Indeed one of Damanhur's symbols is a flower, the dandelion.

Above: Painters at work. Right: A child connecting with a plant.

The Music of Plants

The Music of the Plants research began at Damanhur in 1976, when resident researchers created an instrument that was able to capture the electromagnetic variations of the surface of plant leaves and roots, and turn them into sounds. The desire for deep contact with nature has also inspired the 'Plant Concerts', where musicians perform, accompanied by melodies created by trees. The trees learn to control their electrical emissions, so they can modulate the notes, as if they are aware of the music they are producing. This research has continued, and today, the device used for concerts is available to the public, so that this profound experience of plant world communication may be shared by anyone who wishes to do so.

At Damanhur, we believe that our planet is a living being to be respected and protected, as the concept of 'Gaia' expresses so well. In addition to cultivating a respect for nature, our

ecological vision includes an awareness of how all of humanity is deeply connected with everything that surrounds us. These principles inspire the lives of Damanhurians, and they translate them into practical, everyday actions from renewable energy to organic farming and green building construction, to the community's school system where children are taught to take care of the environment.

The Temples

Damanhur is known worldwide for the Temples of Humankind, an underground work of art dedicated to harmony and beauty as an expression of free

One of the Temples of Humankind.

spirituality. They are a complex of eight halls adorned with remarkable paintings, mosaics, sculptures and glass art—all created to celebrate universal spirituality. The Temples were made with the ingenuity and creativity of Damanhurians, who worked together to excavate the rock of a mountain entirely by hand. It is an 'impossible' dream that awes and inspires thousands of visitors every year. The Temples are a stunning example of sacred and secular art coming together to resonate with values that are important to the whole of humanity.

A Federation of Communities

Damanhur's social structure and political system have been changed many times over the years, from the first communities to the present Federation of Communities. The Damanhurian decision-making model has evolved,

creating an efficient, democratic system, with representatives and elected bodies based on the participation of all citizens in public debate. Changes to the rules and regulations are ratified in accordance with the Constitution, which has been updated several times.

The aims of Damanhur are:
- The freedom and re-awakening of the human being as a divine, spiritual and material principle.
- The creation of a self-sustaining model of life based on ethical principles of good communal living and love.
- The harmonious integration and cooperation with all the forces linked to the evolution of humankind.

The Constitution, the ethical charter of Damanhur, regulates the Social Body,

Different Communities in Damanhur.

which is formed by Damanhur's citizens. The Constitution indicates the principles and aims of Damanhur and represents the red thread that runs through life here. There is a public ceremony on entering citizenship, when the new citizen makes a commitment to respect and apply the norms of the Constitution. There are different forms of citizenship, according to the choice and commitment of the individual. The Communities that are organized as a Federation represent the social network and are inspired by principles of solidarity and sharing. It is also possible for groups belonging to different schools of thought to affiliate with the Federation, if they are inspired by the same aims.

Out of the creation of shared tradition, culture, history and ethics, the popolo is born

The spiritual path, which we call the School of Meditation, leads every citizen through a lifelong process of self-exploration and search for the meaning of existence. This is facilitated through the study of ancient magical traditions and the celebration of the rhythms of nature. On this path, everyone learns to develop their talents and overcome their weak points.

In 2005, Damanhur received recognition from the United Nations Global Forum on Human Settlements as a model for a sustainable society. It was the result of Damanhur's deep respect for the environment as a conscious, sensitive entity, and our citizens' commitment to co-existing with the plant and animal worlds in a reverent and nurturing way. The choice to cultivate organic food and animal husbandry is a natural consequence of a deep respect for life, as is building according to green principles and using renewable energies. Some citizens have even created companies in such fields as renewable energy, eco-clothing and food production. Damanhurians prefer natural healing methods and a holistic approach to wellness, but not to the exclusion of science and medicine. It's about appreciating life in all its forms and having the lowest possible impact on the environment. Where appropriate, cutting-edge technologies are used as a valuable ally in the defence of health and nature.

Damanhur is communal property and all citizens participate in the management of it. Each Community is like a family, where those who work participate in covering costs supporting those in need, giving the capacity of

People of Damanhur.

supporting the projects of the Federation, such as the Damanhur School and the Temples of Humankind. Damanhurians created the Atalji housing association in order to have common ownership of all of the Federation's property assets, both land and the Communities' houses. This economic legacy has grown over time as Atalji has invested in acquiring new land and renovating old houses. Every citizen is a shareholder in the association and can withdraw their shares if they decide to leave. In addition, a cooperative called Punto Verde owns the facilities and land that is used for the community's agricultural activities and livestock.

Damanhur has its own complementary currency called Credito as a new form of economics that is based on the values of cooperation and solidarity. Credito means credit and reminds us that money is a tool through which we grant trust, and represents a return to the original meaning of money as a means to facilitate exchange, based on agreements between those involved. In technical terms, the Credito is a functional account unit, active in a predetermined and predefined circuit. Today, the Credito has

the same value as the Euro. All economic activities present in Damanhur provide for and favour the circulation of Credito as a system of internal exchange. Upon arrival at Damanhur, it is possible for all guests and friends to convert Euro currency.

Damanhurians have created businesses in the fields of arts and crafts, design, construction, renewable energy, nutrition, publishing and more, and many of these are located at the Crea centre. The common denominator among all Damanhur activities is the vision of work as a means of spiritual refinement. Work is a way to offer yourself to others, a way of expressing the creative dimension, and an opportunity to choose materials and processes with a low environmental impact.

In Damanhur, spirituality and community are completely linked. Community is a fractal of the wholeness of being, making each action a manifestation of spirituality.

Location: Piedmont, Italy.
Established: 1975
Area: 500 ha (1,235 ac)
Population: About 1000 people, 600 residents plus 400 people closely connected.
Housing: 25 nucleos = big houses, groups of houses
Common facilities: The Temples of Humankind, Sacred Wood Temple, Damanhur Crea meeting centre, guesthouse, coffee shop, sheds and barn.

Top to bottom: Connecting with a tree; Damanhur's currency, the Credito; A map of Damanhur.

SEKEM Egypt
The desert is alive

by Christine Arlt & Christina Büns

Flowering fields, cotton and date plantations in the middle of what is mostly associated with drought and death— the desert. This is not a mirage, but Dr. Ibrahim Abouleish's SEKEM initiative, which is committed to sustainable development with a holistic approach.

Born in Egypt in 1937, Dr. Abouleish moved to Austria in 1956. He studied chemistry and medicine at the University of Graz where he also received his Ph.D. He then worked as Head of Division for medical research. In Austria, he learned about the ideas of Rudolf Steiner and the Demeter biodynamic standard. When the 40-year-old pharmacist returned from Europe in 1975, he found his home country in very bad shape. A vision came to his mind:

"Sustainable development towards a future where every human being can unfold his or her individual potential; where mankind is living together in social forms reflecting human dignity; and where all economic activity is conducted in accordance with ecological and ethical principles."

Left: One of SEKEM's pupils during a music lesson.

What happened next sounds rather more insane than sane: Dr. Ibrahim Abouleish deliberately bought 70 ha of desert land near Cairo and not in the fertile Nile valley. The experiment worked! By using Biodynamic agricultural methods to maintain the earth instead of exploiting it, he transformed the sand into the 700 ha oasis of today and named it SEKEM— ancient Egyptian for 'vitality from the sun'.

Egyptian organic pioneer

After building up fertile soil, Dr. Ibrahim Abouleish started to establish the SEKEM Initiative, which today works in the four dimensions of sustainable development: economy, ecology, cultural life and social life. "The concept of SEKEM is strongly influenced by intellectual life, including the philosophy, arts and culture that I got to know in Europe. I experienced that development needs a holistic approach—cultural and social life should receive the same awareness as the economic and legislative points of view. And of course all these dimensions are embedded in ecology, which is the valuable foundation of everything," says SEKEM's founder.

Different companies process the agricultural products and distribute them in a fair way within Egypt and the world. Various schools and a vocational

training centre support the individual development of young people by offering a holistic approach to education. The SEKEM Medical Centre serves more than 100 patients daily from the surrounding villages with professional and comprehensive health care. Farmers all over Egypt benefit from SEKEM, and several million customers all over the world cooperate with SEKEM in a fair way.

Empowering young people

SEKEM serves all its 1500 co-workers regularly with cultural activities and training courses in order to empower them, develop their potential and raise awareness on topics like sustainability. Furthermore, in 2012, the Heliopolis University for Sustainable Development

Dr. Ibrahim Abouleish, founder of SEKEM.

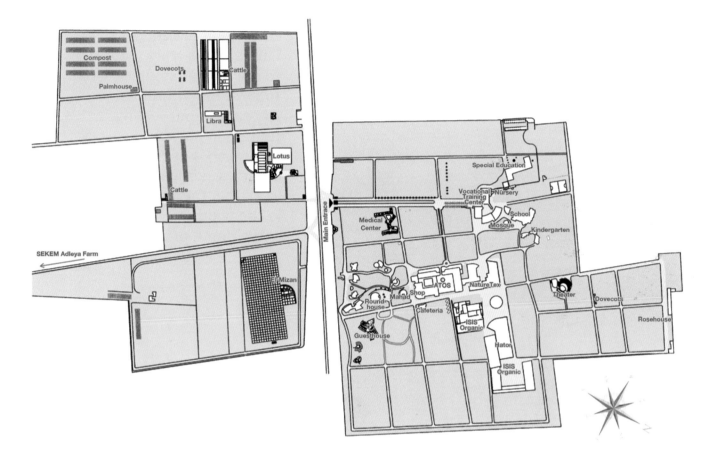

was established under the umbrella of SEKEM. It aims to empower young people to become social entrepreneurs, able to face and overcome tomorrow's challenges through innovation, collaboration and technology.

An economy of love

"When my father returned to Egypt from Austria in 1977, he had a vision, but no experience as a farmer," says Helmy Abouleish, today's CEO of SEKEM. "Everyone tried to talk him out of it. But he proved them wrong: the desert is alive!" SEKEM has been widely praised as an Egyptian organic pioneer. In 2003, the initiative received the Right Livelihood Award, the Alternative Nobel Prize, as a

'Business Model for the 21st Century' and for its 'economy of love'. "We are not just about selling potatoes," Helmy Abouleish explains. "We have created a sustainable community in which people live, work and learn together." SEKEM is a model for sustainable development based on sustainable agriculture—a place where people live and work in a responsible and respectful way, and children learn using a holistic concept in many different fields. SEKEM is a place where all economic activities are done in balance with the cultural, social and ecological dimensions.

Greening the desert.

To wonder at beauty,
Stand guard over truth,
Look up to the noble,
Resolve on the good.
This leads us truly
To purpose in living,
To right in our doing,
To peace in our feeling,
To light in our thinking.
And teaches us trust,
In the working of God,
In all that there is,
In the width of the world,
In the depth of the soul.

by Rudolph Steiner

Location: Egypt, Sharqiyya, 60 km (37 mi) northeast of Cairo
Established: 1977
Area: 684 ha (1,690 ac)
Population: Around 1500 co-workers; approximately 10,000 people from the surrounding villages.
Housing: A huge number of houses and institutions. On the Main Farm, you will find a hotel, several homes and guesthouses.
Common facilities: School buildings, a medical centre, factory buildings, offices and stables.
The SEKEM Head Office is at the campus of Heliopolis University for Sustainable Development near Cairo. SEKEM owns three other farms, two in the Western desert and one on the Sinai Peninsula.

Sekem nursery.

Top: Harvesting dates. Middle: SEKEM Company; harvesting chamomile. Bottom: Calendula field.

Svanholm Denmark
A working and living community

by Andreas Kamp

The Svanholm community started with an advert in a Danish newspaper on May 7th, 1977, with the title: Large Cohousing. The advert was placed by two families that wanted to start a working and living community in the countryside together with others. The invitation to create a large cohousing community with shared economy, common dining, consensus based decision making, equality between professions, and between women, men, children and adults, created enormous interest, and the first meeting attracted 130 people.

The pioneers purchased and took over the Svanholm Estate on May 31st, 1978. Along with the 800-year-old main building and adjoining houses came over 400 hectares (988 acres) of productive farmland and orest. The original intention of working the land without using pesticides became the starting point for organic farming in Denmark, and since 1990, Svanholm has been completely organic. It is mainly a dairy and vegetable farm, but also has sheep, chickens, pigs and goats. Most land is used for fodder, but also for around 80 vegetable varieties, and berry and fruit orchards. Svanholm has sought a balance between a market-oriented and an ideological farming approach. Recent innovations include holistically-planned grazing, a dedicated vegetarian online shop and an experimental forest garden design in the new permaculture settlements.

Starting out with 150 people and only a few habitable rooms in the former estate owner's residence was a practical challenge. Much effort has been put into renovating buildings including establishing housing, common areas and other facilities to accommodate everybody. This development has turned the estate into a modern village providing amenities for individuals, community and visitors.

Above: The 800-year old estate main building.

A dozen pigs satisfy Svanholm's appetite for pork, reduce the amount of kitchen waste and allow for harmonious family pictures.

A holistic life

Svanholm was founded on a dissatisfaction with existing society and the desire to create something new and better. To do that we had to make a fresh start, and so we chose to buy the property. This has given financial restrictions, because of the enormous cost and resulting debt, so we have had to keep our feet on the ground and work hard. Debt or no debt, the people of Svanholm have a large estate of land—a potential paradise, a little enclave—that offers the freedom to try out different strategies for making our dreams come true. Actually, we have developed a kind of state within the State, with our own agreements and ideas. Svanholm has its own internal economic distribution agreements, transition strategy, traffic policy, environmental policy, education policy, health policy and more. In many ways, Svanholm has pioneered developments that have since been implemented in society at large. For instance, we have had different kinds of sabbaticals, long before the politicians in Denmark thought of it. We developed waste sorting systems and worked towards a sustainable way of life, long before the Brundtland report and the introduction of the term 'sustainability'.

Self government

People don't move into the Svanholm Collective to join a specific ideology. We share common goals, but there is not a single common political or religious/ spiritual basis that connects us. Our self-government approach is a way of stimulating people to be more involved in decision-making and to feel responsible for the outcome. We wanted and still want direct influence on decisions that concern our own lives. We are aware that majority rule sometimes can be at the expense of the minority. The common meeting, held once a month, is our decision-making body. We never vote, but negotiate to achieve consensus. We have no formal leaders or '-isms', our ambition is to maintain and improve our sense of community and at the same time accept individual differences—to practise equality without uniformity. This provides reassurance and a sense of belonging, and is at the same time a continuous education in compromise, openness, and reflection about personal versus communal needs and desires.
It is a guiding principle for social

Every year on the 31st of May, Svanholm celebrates its birthday with a communal breakfast under the walnut trees.

and environmental sustainability to manage our use of resources so that we don't use more than we give. How to practise this principle, however, is more complicated than agreeing on it! Practising a sustainable lifestyle has required continuous adjustments in the almost 40 years since the founding of the community. However, external reports show that we have high biological diversity and only one third of the CO_2 emissions of an average Dane.

Our common economy is a cornerstone in our attempt to maintain social sustainability. It allows us to give priority to employment in the community's kitchen, administration, maintenance group and farming and self-sufficiency activities. The increased amount of free time this gives is very attractive to people who want an alternative to the 40 hours of paid work a week, plus 20–30 hours of household chores, of a typical Danish adult. There are many ideological, ecological

and economic reasons for using Svanholm's natural resources to support self-sufficiency. The food such as vegetables, meat, dairy products and eggs that the farm supplies, in addition to what is sold, contribute about half of the community's diet. This balances the desire for homegrown food with the desire for a varied diet. Since the late eighties, two wind turbines produce enough power to cover the community's needs. We heat our buildings with wood chips from our forest and have our own water supply and sewage treatment system. Regarding self-sufficiency, there are, however, still significant gaps to close with respect to animal feed, fertilizer, transport fuels and a popular diet. We have realized that our self-sufficiency initiatives must proceed at a pace that is compatible with developments around us in order to be successful.

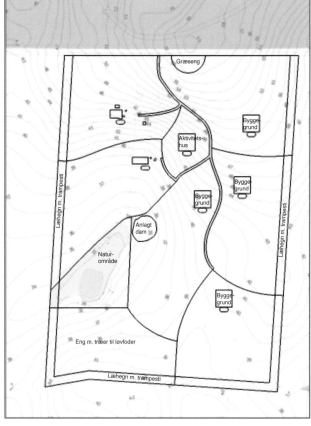

Establishment of the first permaculture settlement at Svanholm in 2016—view to the south.

Location: 50 km west of Copenhagen, Denmark.
Established: 1978
Area: 400 ha (1000 ac). Half of it farmland, the rest woods and park/garden.
Population: 80 adults and 55 children
Housing: 6200 m² (66736 ft²), 14000 m² (151000 ft²) farm buildings.
Common facilities: Power, heat and water supply. Dining hall, large kitchen, ball room, playgrounds, forests and lakes, e-bikes and cars.

Map of the six farms planned in the permaculture settlement.

Aerial photo of the estate's buildings.

The Permaculture Farms at Svanholm

by Esben Schultz & Mira Illeris

In 2016, a group at Svanholm started establishing the first two permaculture farms there. Two young families with children are starting up the project. Each farm is 1.5 ha. The full project is expected to be six farms on 11 ha of land. The farms are on farmland that the project rents from the Svanholm community. The area is on a south sloping hill near the forest at Svanholm. The new farms are converting fields into forest gardens or agroforestry, and are designed to support an almost entirely self-sufficient lifestyle.

The Svanholm permaculture project is remarkable in the sense that the municipality has given it dispensation from relatively strict zoning rules. The reason for allowing the experiment is that it demonstrates environmentally correct farming practices, including high levels of self-sufficiency and biodiversity, independence from fossil fuels, minimal CO_2 emissions and carbon storage. The families will have to meet 75% of their household needs from a land-based livelihood, with no fossil fuels used after establishing the project.

The farmhouses will be small, one-storey buildings, built mainly with local materials like timber, straw bales, clay and stones, and will have green roofs. It is possible

to volunteer at Svanholm and in the establishment of the permaculture settlement.

An existing forest garden used for teaching.

Huehuecoyotl Mexico
Ecology is art

by Giovanni Ciarlo

After fourteen years living as a nomadic tribe of roving artists, we began building our ecovillage in 1982. During our travels, we learned to put on theatrical productions and took on mysterious names like The Hathi Babas Transit Ashram, and then The Illuminated Elephants, a traveling theatre community. Members who joined us later came from diverse activist and youth groups, or were attracted to our progressive values and way of life. After several years on the road, we decided to look for a place to settle. We found our home in the majestic mountains of Tepoztlan, Mexico. We named it Huehuecoyotl, after the Aztec god of poetry, music, dance, and mischief. This was actually the name the property had before we bought it. It is the custom in the region to give every parcel of land a traditional name. Huehucoyotl is surrounded by the Tepoztecan National Forest on three sides, making it part of the forest while remaining private and communal. It is near a steep vertical mountain, and in the rainy season enjoys abundant waterfalls dropping 60 metres down the mountain.

Left: Huehue Family and the Sacred Amate Tree.

To begin with, old converted school buses and vans served as our homes. In one of them we shared a communal office, kitchen and a studio area for preparing our shows. We implemented a shared economy and paid all expenses collectively. We also honoured the Native American traditions. One day an old and wise Klamath elder performed a ceremony for us, in which he named our group Kilokaga Nx Nilaxi, which means 'a small but powerful tribe looking for knowledge'.

All the houses are self-designed and built by the owners. Materials used are local wood, clay, adobe, volcanic stone, brick, ferro-cement and cob. The architecture is vernacular, and integrated with the natural shapes of the mountains. Water is harvested from rainfall and stored in cisterns, and then distributed by gravity to all the houses and recycled. Many of the houses have dry toilets or a combination of dry and flush toilets. There are often houses or rooms for rent on a seasonal basis.

Today, some of us practise a scaled-down version of nomadism. A small group caravanned for 13 years throughout Central and South America, all the way to Tierra del Fuego, until 2009. Other members continue to travel extensively,

Location: Municipality of Tepoztlan, State of Morelos, 120 km (75 mi) south of Mexico City, 2000 m (6560 ft) altitude.
Established: 1982
Area: 3.5 ha. 50% of the land is multi-use, 40% residential, 10% forest.
Population: 22 residents, from newborn to 70. An equal number of men and women. Many nationalities in the group: Mexican, Spanish, Swedish, US, Italian, Danish, Basque, Dominican Republic, and Ecuador.
Housing: 14 single homes and one duplex.
Common facilities: The common house, El Teatro (The Theater), with open space and a kitchen for events of up to 200 people. Dormitory for 30 individuals. Medium size vegetable garden for residents' use.

Above: Communal house and dormitory. Left: Garden sculpture. Below: A plan of Huehuecoyotl.

Natural architecture.

taking bilingual performances to all kinds of audiences. Others choose to work on aspects of the arts, such as theatre, music, writing, poetry, painting, holistic healing, photography, film and video, gardening, crafts, architecture and all forms of personal development. Some have written books about our adventures, recorded CDs featuring our music, printed postcards featuring Mayan calendars, produced videos presenting our lifestyle and developed a wide variety of arts and crafts produced and sold in our ecovillage. The rights of Mother Earth are promoted by our community members as an important step towards regenerating the ecological balance of our planet. Our motto is 'La Ecología es Arte' (Ecology Is Art).

Our daily activities, apart from the diversity of ways we earn our living, include the care of our physical, mental, emotional and spiritual health. We also provide environmental, cultural and wellness training through events, workshops, festivals, retreats, conferences, Eco-tours, audiovisual performances and crafts production. This offers visiting groups and individuals an opportunity to explore sustainable lifestyles, healing methods and nonreligious spiritual practices. We maintain a close and flexible relationship with various neighbouring communities. Our ecovillage is located 1.5 km from Santo Domingo Ocotitlan, a Nahuatl-speaking village of indigenous farmers.

Design students with map.

Over the years, we have collaborated closely with the municipality of Tepoztlan, a traditional indigenous name dating back to Aztec times. We give theatrical performances, organize a yearly dance festival, welcome indigenous groups such as the Zapatistas of the southern state of Chiapas, support local farmers and the local organic market, and participate in local culture. Like Ocotitlan, Tepoztlan is an ancient Nahuat village in the foothills of the majestic, bio-diverse Chichinautzin Corridor, and it has become a weekend tourist destination for city dwellers. Today, many Tepoztecans devote their commercial efforts to the successful promotion of 'esoteric' tourism, fine traditional restaurants, and arts & crafts in its many and varied forms. In the past, we helped to form a food co-op, an alternative school, a crafts co-operative, a sewing collective and other social organizations in neighbouring towns.

Huehuecoyotl has become an experimental showcase of appropriate technologies, participatory decision-making, group facilitation, and other useful tools designed to promote a new type of consensual democracy. Permaculture, bio-regionalism, sustainable architecture, Native American studies, experimental theatre, light and sound performance, the practice of traditional and emerging ceremonies, media, communications and helping to reconstruct the social fabric of society have become the focus of our small but enduring ecovillage vision.

Top: Children creating. Above: Vertical hydroponic design. Left: An early Huehue logo.

Céu do Mapiá Brazil
New life, new world, new people, new earth

by Felipe Simas and Ana Carolina Simas

The purpose of Céu do Mapiá Village (CMV) is spiritual development. It is a healing centre based on the principles of harmony, love, truth, justice and peace, and the ecological principles of conservation and reforestation. CMV is guided by a deep understanding of living spirituality, which recognizes and celebrates the divine in the forces and presences of nature—the sun, moon and stars, earth, wind and sea.

The 'glue' of CMV is the ritual use of Santo Daime, a Christian doctrine based on the ritual use of Ayahuasca. This is a brew made of leaves from a special bush and liana, native to the Amazon forest, that has traditionally been used by indigenous people in this part of the Amazon forest. The Santo Daime Doctrine unites elements of Amazonian shamanism, Christian faith, European esoteric knowledge and Afro-Brazilian culture. It is open, inclusive and eclectic, and represents a peaceful synthesis of many cultural matrixes. Santo Daime spiritual work aims for self-knowledge and the experience of God or the Higher Self within, and promotes the re-enchantment of nature. Scientific research shows that the use of Ayahuasca can be a healing experience that leads to a higher level of physical and mental wellbeing.

To live on natural resources, grow and be together

The history of CMV dates back to the early 70s. A rubber tapper named Sebastião Mota de Melo (Padrinho Sebastião) and his family were followers of Mestre Raimundo Irineu Serra, the founder of the Santo Daime spiritual doctrine in the early 1930s. As Padrinho Sebastião and his family began to hold spiritual sessions, a group of families formed, gathering small farmers, rubber tappers, people from the city of Rio Branco and from other parts of Brazil. Following Padrinho Sebastião's spiritual guidance, the group decided to create a community to daily practise the spiritual teachings delivered by the Santo Daime. Padrinho also warned that the Western development model was an illusion from which we need to free ourselves in order to progress towards a new society truly guided by spiritual principles. The group decided to move to the middle of an isolated forest area, to live, work and pray together, living simply from the forest and agriculture. The idea was to experience human and spiritual development in a new, fair, communitarian lifestyle, in harmony with, and nourished by the forest, with experimental environmental, socioeconomic and cultural solutions. In 1983, the group of 60 families arrived

in today's location. They lived in a communitarian system, working, planting, harvesting and celebrating together. Mapiá became the world capital of the Santo Daime spiritual movement, which has expanded rapidly in the last 20 years, with groups and followers in many countries. During this process, more people joined the community.

With today's population of about 600, CMV has made the transition from a radical communal lifestyle and production, with no money, no television and no telecommunications, to the current international village model. Now each person is responsible for their own income, and we have Internet, telephones and televisions. Twice a year, spiritual festivals bring over 200 visitors to the Village. Transport, ecotourism, accommodation, food, general goods and services are some of the activities that generate income. Some people have pensions and other sources of income

from outside the forest, while some have formal jobs in Mapiá, such as the schoolteachers.

The forest is an important source of building materials (wood, palm trees) medicines and food, for example. Local and regional products come from the forest, local family scale agriculture (manioc, banana, rice, beans and corn), fishing, hunting and animal rearing. There are also local shops that sell industrialized goods brought by boat from the city. Food production is based on natural practices and local resources, with no use of industrialized fertilizers and pesticides. Transition from traditional slash and burn cropping to agroforestry systems started over 10 years ago, and there are several food forests in different stages and with different species.

Céu do Mapiá Village is not connected to the grid. Solar panels generate electricity to over 50% of the houses.

Diesel generators are also used in many houses. Energy is needed also to pump water from wells or from the river to the houses. Alternative solutions to reduce fossil fuel dependency, and at the same time increase energy generation, are necessary to support community sustainability. Rainwater harvesting systems are common in most houses as a way to reduce the use of generators.

"I was born and raised in the Forest, and here I stay, and I don't want to leave it, not at all! It was where I met my Eternal Life, which is ever present. Eternal life is in everything, and transforms what is dead to Life. I pray for more love and perfection for people every day. "

Sebastião Mota de Melo

Founded in 1987, the Dwellers Association is responsible for the overall organization of the community. It is structured in several departments (social assistance, human health, residue management, food production and disciplinary council). Every Monday is the community´s common workday, when all dwellers are expected to work together on the maintenance of common spaces and buildings. The Santo Daime church has a central role in the village governance, bringing the community together to practise and study

the teachings of Santo Daime. It is a major influence on the organization of the community, and in all aspects of daily life. The main spiritual and community leaders today are Sebastião Mota's wife and son, Rita Gregório and Alfredo Gregório de Melo. A spiritual council of elders plays an important role in decisions regarding the community.

As the village develops and grows, several other institutions and groups have been formed in a continuous process of community organization: the Forest

Medicine Centre and others. CMV has a higher living standard than most of the settlements in this region, which are isolated and without official help.
It is a reference for wellbeing, health and education in an extremely isolated and socially vulnerable part of Brazil.
In 2004, a Community Development Plan was created and resulted in the strengthening of groups and organizations and established the Interinstitutional Work Group. Although many decisions are still made by community leaders, elders and other dwellers in a spontaneous way, this group is today the higher council for deciding community issues, representing everyone.

In 2013, the community hosted the AmaGaia Ecovillage Design Education in the Amazon, certified by Gaia Education. Members from all community groups, school students, representatives of local and national institutions, people from other communities in the region, indigenous peoples, visitors and spiritual leaders got together for 30 days with experts in ecovillage design from Brazil and abroad. Based on real case studies, the participants formed workgroups that co-created projects to improve community life, and make it more sustainable in the social, economic, ecological and spiritual dimensions of community living.

Ceu do Mapia is without doubt a special ecovillage experience. It is isolated in the Amazon forest, has been founded by traditional people, gathers dwellers and visitors from all over the world, and is a keeper of traditional knowledge regarding life with and from the forest, in particular the ritual use of Ayahuasca. We make continuous efforts to implement community projects that strengthen local sustainability.

Location: Floresta Nacional do Purus, Amazonas, Brazil.
Established: 1983
Area: 120 ha. In the Purus National Forest, an Environmentally Protected Area of Amazon forest with mega biodiversity.
Population: Ca. 600 people. Housing: ca. 130 homes, mostly single family houses.
Common facilities: Church, Feitio house (for the ritual of Ayahuasca production), Healing Centre, the Star Temple, school, community kitchen, Dwellers Association, Forest Medicine Centre, Nature Garden Crafts School, Cooperar (food production and distribution cooperative), Environmental Health Group, Music House, Telecentre, a small library, a local radio station, Crafts House, Youth Cultural Centre, agroforestry systems, annual cropping areas, manioc flour house, children's playground, soccer field, small markets and a graveyard.

Kibbutz Lotan Israel
An oasis in the desert

by Alex Cicelsky

The long journey through the desert to Kibbutz Lotan is stunning as the road meanders through millennia-old rock formations. We live in one of the hottest and driest deserts on earth. The physical site that was chosen was secondary to the common vision we held, of building a community where we would work together for the common good, and would have equal opportunity to realize our potential. We had a lot of time to talk about our ideals, because it took four years from when we started as a group until we moved into our houses. That long wait was probably good considering we were all 18 years old when the decision was made to start the kibbutz!

When I give tours I point out a number of things that are not obvious, as we walk on tree shaded paths, past experimental solar-powered straw bale houses. The first is that we did not start as an ecovillage; we began as an intentional socialistic community with egalitarian Jewish values, sponsored by the 100-year-old Kibbutz Movement. Kibbutzim were built in cooperation with the Israeli government to establish agriculture and industry in the country's periphery. Most land in Israel is owned collectively by its citizens, (as are all water resources) so we pay rent. Our houses, buildings, infrastructure and agricultural machinery were supplied by the government on long-term loans, which we repay. We, the young adults that founded the community, chose to be members of the commune where we'd all work the land, share our income (and pay our debts) and develop a spiritually motivated society. What drew me into this adventure was both the physical and ideological aspects. With the guidance of advisors from neighbouring Kibbutzim, we changed from city dwellers to hard working farmers, carrying box upon box of vegetables out of our fields of pure sand. We harvested dates from the trees we planted and everything was irrigated with brackish water. The work in establishing our economic base from agriculture was physically very hard, and not always economically successful. The 'back to the land' ethic is measured in its ability to pay at least for food, water, health care, housing, supplies and machinery. We spent many hours in committee meetings, and at-large community meetings discussing the income of our communal business. We also spent considerable time considering the values of our society.

Our community was unique in Israel because we decided to meld the rather strict and secular structure of socialism, with the egalitarian and pluralistic interpretation of Liberal-Reform Judaism.

Green Apprenticeship Permaculture course students.

Caring for people and appreciating nature

Most of us came from Jewish homes with various levels of religious practice, so we all had unique cultural gifts to bring to the group. We celebrate Shabbat (the Sabbath) every week and holidays throughout the year with creative prayer, music, foods, and a diversity of traditions from our families and their heritages from around the world. We put great effort into making holidays fun and meaningful. And we make a concerted effort so that Shabbat and holidays are time for physical rest and spiritual renewal. Shabbat and holidays begin at sunset and end the next day when three stars appear in the night sky. During this time there is no formal work, with the exception of milking and feeding the cows, and community food service in the dining hall. In addition to not working it is taboo to speak about making money or labour. The result is calmness and time for

connecting with each other and nature. As our knowledge of farming grew, so did our understanding of the connections between agriculture and the seasonal patterns of the Jewish holidays. We deliberately incorporated these connections into our holiday celebrations, making the celebrations much more meaningful to the lives we were living. Judaism is based on commandments of caring for people and appreciating nature. Combining a farming life that depends on nature with Jewish values had a profound impact on those of us that had studied ecology, clean technologies and organic agriculture. This, for me, was a paradigm that catalysed a need to add the dimension of 'eco' into our village.

To our community's credit we always valued sharing of resources and had from the beginning many ecovillage practices. Our few cars are collectively owned and to use a car I have to sign up on a board in the office, including where I'm going,

Eden's bat mitzvah.

so that others can join me. The communal dining hall serves three healthy meals a day, meat is served only five times a week. Almost all of the food served in the dining hall is "home made" from locally sourced produce. All of our houses are small and children play all over without regard to whose property it is. We rotate positions of leadership and have a direct democracy, with long processes to reach close-to-consensus decisions. We live frugally both by choice and because we have very limited income. We love the desert around us and work to protect it.

An educational demonstration center

In 1996 a few of us decided to start work on a research, development and education centre that would address a wide gamut of environmental issues and would affect change in our community and throughout the country.

Not only was the scope of the change we envisioned daunting, but the resistance to change from our fellow community members was significant. Many community members were concerned that these changes would demand more work (we're always understaffed) and investments that we could not afford. One issue was that none of our pre-fab concrete houses were appropriate for our climate. In the winter the walls of the buildings are 12°C and during the 5 month long summer the air temperature is 30–45°C. There was no recycling at all in the country. Conventional agriculture paid the bills. Nevertheless we set our sights on what we felt needed to change.

We spent four years learning, taking our first actions and making educational mistakes. Our list of changes included community-wide separating and recycling of waste, developing organic and chemical free agriculture, developing energy-efficient building

systems based on natural materials, developing renewable energy production, reducing water usage and increasing wastewater recycling, protecting migratory birds, and starting an educational demonstration centre so that we can change our region and the country. Slowly the community came to love what we were doing, and wanted to participate in the projects. In 2000 we established the Centre for Creative Ecology (CfCE) which has had a tremendous impact on the community and the region. Our work was recognized and applauded by our Regional Authority, the Ministry of Agriculture and the Ministry for Environmental Protection for voluntarily taking action to reduce our waste by over 70% and for serving as an inspiration to other communities in the region and the country. We were in the news because we composted our organic waste and turned the refuse that doesn't decompose into a Noah's Ark playground, bird watching hides and the community bus stop. The creative work was done by community members, mud-covered

kids and participants in our Green Apprenticeship Permaculture Design Course. This recognition and support inspired us to continue our research, development and education work.

The most magical place on Kibbutz Lotan is where we teach organic farming, soil care, nature conservation and sustainable development. The organic garden, Israel's first CSA (Community Supported Agriculture), is nothing less than a miracle. The southern Arava desert is one of the hottest and driest on the planet. Growing organic vegetables in pure sand, without industrial fertilizers, pesticides or herbicides, was considered pure folly by agricultural advisors. Years of building soil and using composted waste from the communal kitchen and the dairy, has created a bountiful garden. A large and diverse collection of vegetables, herbs, fruit trees and flowers are home to increasing populations of insects, butterflies and bees. Each year,

Tu B'shvat creative service.

Harvesting vegetables.

birdwatchers from around the world flock to the garden, which is a resting and feeding place for hundreds of varieties of migratory birds each fall and spring.

Natural building

The Eco Campus is the solar-powered neighbourhood where our students live and learn. It has 10 earth plastered straw bale 'dome-atories'. These experimental buildings, constructed by the students, are earthquake resistant due to an interior, handcrafted geodesic steel pipe frame, and may be the most energy efficient buildings in the country. They use about 70% less energy to heat and cool than conventional buildings. These prototypes are teaching us how we can retrofit existing buildings and build new houses for new members of our community.

Outreach, education and eco-tourism

EcoKef ('kef' is 'fun' in Hebrew and Arabic) is our hands-on learning park

where thousands of visitors each year see the feasibility of sustainable desert agriculture, natural construction and water treatment and renewable energy options. The sole acacia tree that lived here before we arrived is now surrounded by olives, figs, marula, moringa, and thousands of date palms producing what are considered some of the best Medjool dates in the world. The water used for all the trees is brackish ground water unfit for human consumption. We also have an orchard irrigated with treated wastewater from the city of Eilat. We ran the first EDEs (Ecovillage Design Education courses) in the world. In addition to our Permaculture courses, we host the southern branch of the Arava International Centre for Agriculture Training. Fifty college students from southeast Asia get weekly instruction in advanced agriculture and Permaculture, and we run seminars for Israel's Agency for International Development Cooperation and for university groups from around the world. Students and visitors of the CfCE have the

Sheet mulched garden doing well.

opportunity to meld with the Kibbutz Lotan community and have a first-hand experience of intentional community life, community celebrations and engagement in protecting the environment. We have renovated 22 apartments into guest accommodation. Our strawbale Solar Organic Tea House is the first grid-connected solar restaurant in the country and has no-complaint composting toilets.

Spirituality, community celebration

Kibbutz Lotan is one of the first communities in Israel where liberal, egalitarian Judaism is practised. I think that ours is 'a Jewish expression of the Universal belief' of spirituality, social justice and earth-care. Our ceremonies are full of song and poetry which resonate with people of different faiths and spiritual practises.

Doing beautiful things together

I love Tu B'shvat—the New Year of the Trees—which is the day in the early spring that was used in ancient times for calculating tithes. Today it is a nationwide day of planting trees. When driving home I stop at the overlook and think about all of the years of planting these trees that has resulted in Lotan appearing like a green island in a sea of sand. I think our community's favourite holiday is the Fall Harvest Festival of Sukkot when we spend the week living outdoors in makeshift huts with roofs of palm branches. It is a tradition from Biblical times to remember the travels out of Egypt and slavery, through the desert and into the land of freedom and abundance. Today we see it is a reminder that we are sojourners, temporarily on this Earth and dependent upon her. It is with this sense of responsibility to the Earth that we balance the development of our community, our use of resources and protecting the environment around us. It is also an opportunity to do beautiful things together.

Location: In the southern Arava valley of Israel and Jordan, Southern Israel.

Established: 1983

Area: 62 ha (153 ac) zoned for residential and commercial use. In addition, there are 178 ha (440 ac) of fields and date palm plantations.

Population: In total 200 people. With 30 families of members: 60 adults and 60 children. In addition: families of refugees, 18-year-old Israelis in a 'year of service and leadership', volunteers from around the world, Centre for Creative Ecology interns, 15-20 Green Apprenticeship Permaculture students.

Common facilities: Community Centre, Seminar Centre, Bar and Cinema, Dance-Hall, workshops, Forest kindergarden, farm and more.

Dome construction and clay building.
Top right: Aerial photo of the EcoKef.

Torri Superiore Italy
An ecovillage from the Middle Ages

by Lucilla Borio

The village of Torri is mentioned for the first time in a document dated 1073. The origin of the medieval settlement—a complex of dwellings separated by a few hundred yards from the main village—is uncertain, though it may date from the late 13th century. This was a time of great social and religious unrest in the region, which explains the village's architecture, a stronghold remarkable for its width and height. The village covers a total area of 50 by 30 meters, on eight levels, with a floor space of close to 3000 m².—good protection for its inhabitants.

Torri Superiore is notable not only for its compact architecture, but also for its good state of preservation. The medieval village consists of three main buildings, separated by two partially covered inner alleys. An intricate labyrinth of stairways and terraces link 160 rooms with vaulted ceilings. The stone, lime and sand used was of local origin, from the surrounding valley or the nearby riverbed. The buildings have been added to over the centuries, with the last parts of the hamlet probably built around the end of the 18th century. This was when the village reached its highest population, before gradually being abandoned as people began to leave Liguria due to lack of employment. Traces remain of communal living: a large hall used possibly as a common kitchen, an open-air oven and an intricate, close-knit pattern of rooms and terraces that create unexpected and surprising views at every turn. During the course of the 20th century, Torri Superiore was abandoned to decay, becoming a ghost town. Its beautiful towers and productive terraces were deserted and falling into a state of ruin.

In the early 1990s, the Cultural Association began purchasing the village from its then owners, with the aim of restoring it as an ecovillage and creating homes and jobs for a new community of residents. Over the next few years a detailed study of the village's structure took place, before the beginning of a complex restoration programme that balances private and public areas. A refurbishing plan was designed that would preserve and enhance the character of the medieval village through the use of natural, eco-friendly materials, appropriate principles of bio-architecture and by working in harmony with the surrounding environment. Wherever possible, local firms, supported by members, residents and volunteers from all over the world, have helped with the building work.

The restoration started in 1996 and by 2015 all the communal areas were

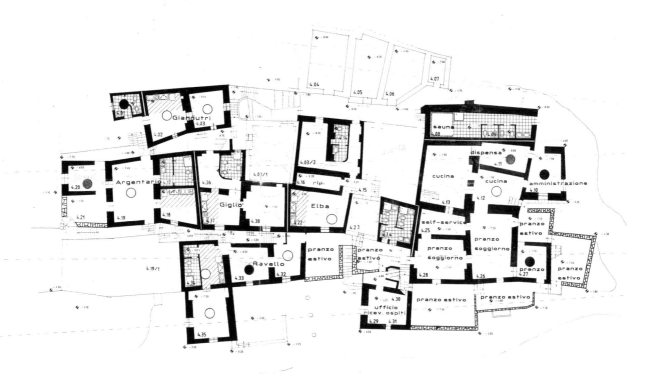

Mapping of the village in the early stages of the reconstruction, 1991.

completed, including the cultural centre, and 21 residential units of the 22 planned. Solar panels for the production of hot water were installed, and photovoltaic panels for electricity. There are low-temperature heating systems in the guesthouse and many private houses (max 18°C). Our electricity is mostly solar, or supplied by a green company. For restoration work, local stone, natural lime and natural insulation (cork and coconut fibre panels) are used. All our doors and windows are sustainable wood, and we use only eco-friendly paints. We have an outdoor compost toilet, and an example of natural water treatment for a private residential unit.

We still have many plans to make the ecovillage better, like more solar panels, including photovoltaic ones for electricity. We want to increase the amount of food we produce using permaculture methods, and generate less waste by closing more of our 'cycles'. Since 2010 we organize 'Living in Transition' workshops on

sustainability and eco-friendly living, to pass on some of the skills we've acquired over the years.

Sustainability in practice

Torri Superiore is the ecovillage, which includes all the resident and non-residential members, and guests that join in. The idea of restoring the village was always based on using eco-friendly principles. Being involved in the GEN Global Ecovillage Network, in the Italian ecovillage network RIVE, and in the permaculture movement, has helped the group to develop and realize many of the practical goals in a more sustainable way.

There are organic permaculture gardens and fruit orchards that provide fresh produce, with more on the way, and free-range chickens that provide quality eggs. Homemade products include bread, fresh pasta, olive oil, honey, jams, yoghurt and ice cream, local herbs for cooking and tea making. The food is

organic and local, or at least national, as much as possible. No frozen foods, genetically modified or prepared products (additives, sauces or dressings) are used, and in winter pizza and bread are baked in wood-burning stoves. All food waste is fed to animals or composted. All other waste materials are separated and composted, reused or recycled as much as possible. The residents own only five cars and encourage guests to use public transport (train and bus). There are two lovely donkeys, used as working animals around the village and for the gardens.

The resident community at Torri Superiore started as a small group in 1993 and currently has 22 permanent members, including three children and five teenagers. This is a group of people from Italy and Germany involved in a common project of rebuilding, education and responsible living. Each family has a private home

with a kitchen, but enjoy sitting down to a meal together, usually in the dining room of the guesthouse.

Residents of the community look after all the day-to-day activities that take place at Torri. They organize events, celebrations, parties and art activities; manage the restoration of the ecovillage; and develop organic farming and permaculture programmes (including the small-scale production of olive oil, vegetables and fruit for selfconsumption); and the housing of chickens and donkeys.

The community meets once a week and makes its decisions by consensus. The common language is Italian. Volunteers are always welcome to the ecovillage, either on a short term (from one to three weeks) or for longer periods through the European Voluntary Service programme.

Location: 11 km (7 mi) inland from the town of Ventimiglia, north Italy.

Established: The Torri Superiore Cultural Association was founded in 1989, but the medieval village is about 700 years old (13th c.). Restoration of the building started in 1996.

Area: The building is ca. 2500 m² (27000 ft²) altogether, on seven levels, with: 1 guesthouse 700 m² (7500 ft²), 22 private homes of ca. 1400 m² (15000 ft²), cellars & storage rooms 400 m² (4300 ft²), plus about 25 terraces. About 3 ha (7.4 ac) of land (mostly olive groves and vegetable gardens).

Population: 22 residents in the ecovillage, 250 residents in the Torri village. Ecovillage residents between 6 and 62 years old.

Common facilities: The guesthouse has dining rooms, kitchen + storage room, library, playroom, bedrooms, bathrooms, meeting room, yoga room, 8 terraces. There are also farm huts in the valley.

Harvesting olives.

Above: Party in the piazzata. Right: The Bevera River and the swimming holes.

EcoVillage Ithaca USA
Showing us the future

by Liz Walker

EcoVillage Ithaca (EVI) is developing an alternative model for suburban living, which provides a satisfying, healthy and socially rich lifestyle, while minimising ecological impacts. It is the largest, and one of the most well-known ecovillages in the U.S. EVI is recognized nationally and internationally for its pioneering work in developing a mainstream, green community with an exceptionally high quality of life, one that appeals to middleclass Americans, while cutting resource use by 60% or more.

We have a comprehensive approach to sustainability and we demonstrate best practices including: green building, densely clustered housing, low energy and water use, strong social ties, local food production, extensive waste reduction, affordability, accessibility, onsite businesses, open space preservation, and hands-on education. Our project is a grassroots development that has created a strong, participatory community.

Over 90% of the EVI land is preserved as open space for farming, wildlife habitat and recreational trails. The land includes meadows, woods, wetlands, streams and ponds. 20 ha (50 ac) of land

Left: Katie in SONG.

have been set aside as a permanent conservation easement and are administered by the Finger Lakes Land Trust. There are three organic farms on the site, which together provide vegetables and fruit for 1,500 people in the greater Ithaca area. The 100 homes of the community are densely clustered on a footprint of just 3.64 ha (9 ac) of land, nestled within a total buffer zone of 6.07 ha (15 ac).

Learn@Ecovillage, the grassroots organization that founded the entire project, continues to work on the long-term vision, and educates the public. It is a project of the non-profit Center for Transformative Action, which is affiliated with Cornell University. Learn@EcoVillage works with students and researchers, often by utilising the 'living laboratory' of the village and small farms. It provides tours and immersion experiences to over 600 visitors annually. Recently we have been training architects, builders and engineers how to design and build Net Zero buildings, which produce more energy than they utilize year-round. Learn@EcoVillage also provides mentoring services and workshops for forming communities.

From vision to reality

In June 1991, Joan Bokaer and I organized

Annual 'Guys Baking Pies' celebration.

an 'Envisioning Retreat' which brought together 100 people from around the US. This Envisioning Retreat discussed and adopted the basic concepts for EVI: a pedestrian village for up to 500 people, made up of cohousing communities, with lots of open space and organic farms, and ongoing educational opportunities. By the end of the five-day retreat people were very excited by the vision and a critical mass of people felt empowered to bring it to life.

We began by securing land. After searching for developable parcels that included good farmland, we chose a 71.2 ha (176 ac) parcel that had been slated for a subdivision before the developer went bankrupt. The prior developer's plan called for using 90% of the site for developing 150 homes, with just 10% open space—a typical suburban subdivision. In our Envisioning Plan, 90% of the entire site is set aside for farms, woods, ponds and meadows, with just 10% used for 100 homes, common houses and parking.

Guidelines for development / site planning process

A committee created a comprehensive plan for long-term land use and development over the course of the first year after we purchased the land in 1992. We adopted a participatory strategy, with public forums that ranged from 60 people working together for an entire weekend, to a final session of just 12 people for an afternoon. The resulting Guidelines for Development have proven to be an important touchstone and set a high standard for both social and environmental sustainability. At the same time a proposed site plan was developed, using the same participatory methods. The final site plan clusters the three neighbourhoods in a triangle with a Village Green in the centre. The Village Green includes a

Porch musicians.

Catherine eating a strawberry.

picnic area, a small pond, a day-lighted stream, and beautiful southern views. It provides a large outdoor gathering space so the entire village can come together for events.

Farming at ecovillage
In 1992, Jen and John Bokaer-Smith started West Haven Farm, with a Community Supported Agriculture (CSA) which has now grown to 250 shareholders on 4.5 ha (11 ac). The farm currently feeds about 1,000 people a week during the growing season. The farm, which is certified organic, grows 250 varieties of vegetables, some fruits, flowers and herbs.

Another successful on-site farm is Kestel's Perch Berries Farm, a no-pesticide, 2.2 ha (5 ac) U-Pick farm with six kinds of berries. This farm is also set up as a CSA, and attracts customers from around the area. Both farms lease land from the non-profit for the cost of the taxes paid on the land.

A third farm is a teaching farm for immigrants who want to create their own small farm businesses. Groundswell Center for Local Food and Farming,

initiated by Learn@EcoVillage, now runs the on-site 3.2 ha (8 ac) Incubator Farm, but also teaches classes at many other local farms.

First neighbourhood
Once the land was purchased, we started organising the First Resident Group, affectionately known as 'FROG', in the summer of 1992. The budding cohousing group spent four and a half years of intensive meetings to plan the neighbourhood, go through a gruelling town approval process, and build the thirty homes and 465 m^2 (5,000 ft^2) common house. The FROG common house is a community centre serving 30 households, with laundry, dining and kitchen facilities, play areas, living room, and eight 'home' offices. FROG home

TREE Common House, with 15 apartments.

Mission
To promote experiential learning about ways of meeting human needs for shelter, food, energy, livelihood and social connectedness that are aligned with the long-term health and viability of Earth and all its inhabitants.

sizes are only 60% of typical US homes and consume 29% of the water and 60% of the energy, and their ecological footprint is 44% of the US average.

Second neighbourhood
In 2001 the first homes were built in SONG (Second Neighbourhood Group), in two phases: 'SONG, Verse 1', with 14 homes and 'Verse 2', with 16 homes, for a total of 30 homes by 2004. The common house was built in 2005–2006. Most homes were built using Structurally Insulated Panels, while some used timber-frame design with straw-bale insulation in the walls. SONG has an extensive free library, and a game room with ping-pong and pool tables for use by the Village.

Third neighbourhood
TREE (Third Residential Ecovillage Experience) decided to use the same kind of standardized design as FROG, thus lowering costs, and used some of the same charming, winding, European street design as FROG, while selecting a middle range of spacing between the houses

compared to the other neighbourhoods. The TREE neighbourhood is one of the greenest neighbourhoods in the US, with all homes reaching the LEED Platinum standard, and some at Net Zero. In addition, the TREE neighbourhood won two awards from the US Department of Energy. The TREE Common House includes 15 apartments, many of which are rentals, allowing lower-income people to be residents.

Guidelines for development, green building, energy efficiency, and renewable energy
All homes are passive solar, superinsulated, and many have photovoltaic panels and solar hot water heating. Passivhaus home design used in TREE reduces energy use by 80–100% compared to typical US homes. The first neighbourhood has installed a 50 kW ground-mounted photovoltaic system, which provides 51% of the electricity for 30 homes. TREE also has a 50 kW solar array. Over half of SONG homes have rooftop solar panels.

FROG Common House.

Densely clustered housing and open space preservation

EVI is a pedestrian village of three neighbourhoods—100 homes on a footprint of less than nine acres. More than 90% of the 71 ha (175 ac) site is set aside for natural areas, farming and wildlife habitat.

Modelling low resource use

There have been ecological footprint studies of EcoVillage Ithaca residents by graduate students from MIT and Cornell University. The latest study was by Jesse Sherry, from Rutgers University, who wrote his 2014 PhD dissertation on the topic. He found that overall, EVI residents reduced their ecological footprint by 64% compared to the US average, and that they used almost two-thirds fewer resources for food, transportation, energy and housing, than typical Americans.

Strong social ties

The three EVI neighbourhoods are based on 'cohousing', with shared common facilities and many shared social events, including several community meals a week. Residents love living at EVI. Many people describe it as an extended family—everyone knows everyone else.

Local food production

Two resident-owned farms supply organic fruits and vegetables to 1,500 county residents during the growing season. In addition, many residents grow their own food in community gardens.

On-site businesses

Almost half (45%) of wage-earning residents work or telecommute from home offices, or provide services for neighbours, lessening the need for commuting.

Extensive composting, recycling and re-use
Residents compost all non-meat kitchen scraps and have cut the need for garbage services by 75%.

Affordable, accessible
TREE was built as affordably as possible, while also putting planning for aging in place. Long-term affordability is enhanced by cutting utility bills to a minimum. One family with four children calculates that they have already saved $5,600 in the first three years of living in their extremely energy efficient TREE home, because their utility bills are so low.

Hands-on education
Learn@EcoVillage works with university students, budding cohousing groups, green building professionals, and anyone interested in learning more about sustainable communities. Our educational philosophy is to engage learners in hands-on practices, as well as to inspire them with a vision of 'another world is possible'.

Public recognition & appeal
From its inception, EcoVillage at Ithaca has enjoyed great recognition as an integrated model of environmental and social sustainability.

It has received local, national and international awards, and has been consistently covered by major national and international media. This public appeal is based on how well these best practices work together to create a deeply satisfying way of life, one that speaks to the need for both social connection and connection with nature. At EVI we have found that the mere physical presence of a community that attempts to lead a life based on environmental and social values is inspiring to many people. There is a common 'aha' experience that one woman described well after a tour. She said: "You are showing us the future."

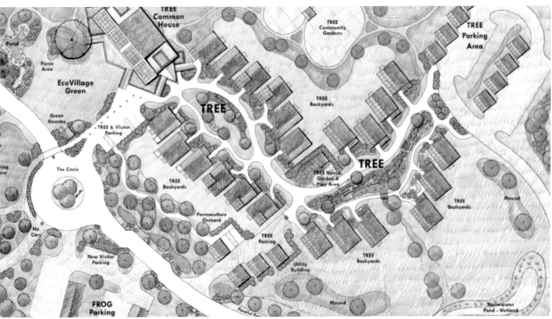

Location: Finger Lakes region, Ithaca, NY, USA.
Established: Planning 1991. Purchased land 1992.
Area: 70 ha (173 ac), of which 4 ha (10 ac) forest, 8.5 ha (21 ac) organic farms, 8000 m^2 (86000 ft^2) of community gardens, three ponds.
Population: 240 people, 75 children, 62 adults over 60. From 1 to 83 years old.
Housing: Three cohousing neighbourhoods, each with a Common House. FROG = First Resident Group, 15 duplexed townhouses (total of 30 homes). SONG =Second Neighbourhood Group, 15 duplexed townhouses (total of 30 homes). TREE = Third Residential Ecovillage Experience, 17 single family homes, 4 duplexes (8 homes), 15 apartments (total of 40 homes).

Lilleoru Estonia
Applying awareness in daily life

by Ave Oit

Lilleoru is a training centre and intentional community, designed and built with an aim to support the conscious and holistic development of a human being.

School of awareness

From the very beginning, the practical application of awareness has been the key element in all activities. Kriya yoga teacher Ingvar Villido and a small group of his students initially established the centre in 1993. Together, the first houses were built with a permaculture garden, and the Flower of Life Park was created. Today 108 people form the core group. They come here regularly to study, take care of the community and place, and organize all necessary activities.

'The Art of Conscious Change' techniques that are applied here have been developed by Ingvar Villido and have reached thousands of people in Estonia as well as abroad.

Lilleoru's spiritual name is Sat Chit Agasthiswarar Gurukkulam.

The Flower of Life Park

At the heart of Lilleoru lies the Flower of Life Park, based on an ancient pattern of creation. The park symbolizes the unity

of human essence—universal wisdom and harmony that belongs to us all. The Flower of Life is beyond any particular religion, belonging to all of humanity. It is a geometric representation of the creative processes of life, the original language of the universe, pure form and proportion. It contains within it the life cycle of the fruit tree: tree—flower—fruit – seed—tree. The five stages in the life cycle of the fruit tree have a parallel in the whole divine process of creation.

SkyEarth ecovillage

The village started because many people wanted to live closer to Lilleoru and the teachings that are shared here. SkyEarth was planned and built by those who study at Lilleoru. Planning started in 2005 and has been a learning process for present and future residents—part of a path of conscious change. This aspect has made the development process of the village unique, compared to the development of typical new contemporary settlements. The social structure of the village, the close cooperation of residents, the economics of the whole process and the choices of the residents—in terms of both building design and materials—make it a consciously created ecovillage.

Left: Flower of Life Park

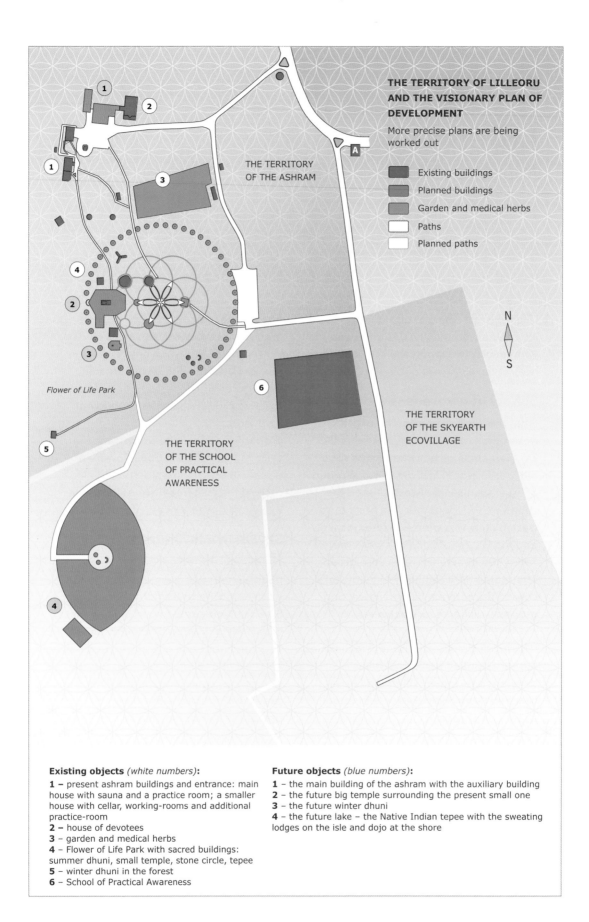

THE TERRITORY OF LILLEORU AND THE VISIONARY PLAN OF DEVELOPMENT

More precise plans are being worked out

- Existing buildings
- Planned buildings
- Garden and medical herbs
- Paths
- Planned paths

THE TERRITORY OF THE ASHRAM

THE TERRITORY OF THE SKYEARTH ECOVILLAGE

THE TERRITORY OF THE SCHOOL OF PRACTICAL AWARENESS

Flower of Life Park

N
S

Existing objects *(white numbers)*:
1 – present ashram buildings and entrance: main house with sauna and a practice room; a smaller house with cellar, working-rooms and additional practice-room
2 – house of devotees
3 – garden and medical herbs
4 – Flower of Life Park with sacred buildings: summer dhuni, small temple, stone circle, tepee
5 – winter dhuni in the forest
6 – School of Practical Awareness

Future objects *(blue numbers)*:
1 – the main building of the ashram with the auxiliary building
2 – the future big temple surrounding the present small one
3 – the future winter dhuni
4 – the future lake – the Native Indian tepee with the sweating lodges on the isle and dojo at the shore

Temple, yantra and the surrounding herb garden..

Herb garden

Our herb garden is located in the heart of Lilleoru, in the Flower of Life Park. Powerful herbs—the ingredients of herbal teas—grow on the petals of the Flower of Life Mandala and on the stepped terraces of the park. Lilleoru herb garden follows the principles of organic agriculture and permaculture. Organically grown food is clean and nutritionally complete, preserving the health of humans and our environment. All Lilleoru tea blends are made from carefully selected plants. The seeds and plants are planted according to nature's rhythms, on the most powerful days, and harvested during the time when the active ingredients are in perfect balance.

After that, the herbs are dried, with special care to preserve their healing properties and beautiful appearance.

Medicinal herbs and herbal blends are a well-known way to maintain and renew your health, and their use in folk medicine dates back to ancient times. Every living thing is filled with spirit, including plants. Working with the land and plants is an important part of activities at Lilleoru. It is a valuable tool for stress reduction and inner development, providing a gradually deepening experience for the practitioner.

The School of Practical Awareness under construction.

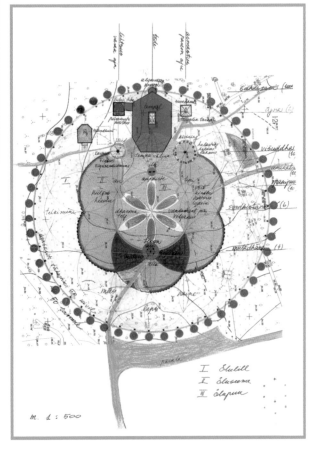

Location: Estonia, 25 km from capital Tallinn.
Established: 1993
Area: 30 ha (74 ac). With forest, agricultural land, water and meadows.
Population: 59 inhabitants including 13 children, aged between 2 to 82 years, live in Lilleoru training centre and SkyEarth ecovillage. 108 members of Lilleoru NGO take care of the community and place.
Housing: One communal living house with 10 private rooms, shared kitchen and bathroom. 3 small cottages, 6 semi-detached houses and 12 single houses.
Communal buildings: Four multifunctional community houses, for meetings, catering, lectures, meditation, courses, office, accommodation and more, temple, dhunis (fire ceremony grounds), Native American tipi, greenhouses, sheds (for gardens, bee keeping and firewood), workshop, summer cottages, shop, composting toilets.

Yantra, the symbol of the five levels.

Tamera Portugal
Paradise under construction

by Leila Dregger

Since 1995, Healing Biotope 1 Tamera in Southern Portugal has been working to create a model that demonstrates how people can live together, not by destroying and exploiting nature, but by caring for her and learning to communicate with all of her beings. The Tamera community of today, 170 members, endeavours to develop concepts of how whole regions can step out of the insanity of globalization. The project was founded by sociologist Dieter Duhm, theologian Sabine Lichtenfels and physicist Charly Rainer Ehrenpreis 40 years ago.

Tamera today

A walk through the Tamera of today takes us through cascades of lakes and ponds, on the shores of which grow the permaculture gardens used for both teaching and food supply. The Solar Testfield demonstrates techniques for cooking and cooling with solar energy and biogas, electricity generation, food preservation, and water pumping. Almost all of the systems were built in Tamera's own workshops. In the research greenhouse, new technologies are tested. Simple traditional construction techniques are combined with modern architectural concepts.

Welcome ceremony around Lake 1 of Tamera.

Tamera is an education centre for the future. Young peace workers come from all over the world to study the knowledge and skills needed to build peace villages and autonomous settlements. The main fields of study are ecology, energy autonomy, and inner and outer peace work. The basis of these studies is learning to build community and gaining social competence. Men and women from different countries and cultures are working together, contributing their knowledge and experience and gaining insights, which they will use for the creation of autonomous settlements in their home countries. The most important aspect in all of this is the coming together. People who previously learned to perceive each other as enemies, for example those from Israel and Palestine, are working side by side. Common work towards a higher goal leaves no space for hostility. Compassion, responsibility for the whole and mutual support are the basic ethical guidelines for living together in Tamera.

In addition to specialist knowledge, participants bring back home the joy of experimentation, greater self-esteem and an experience of community. Research, education and participation unite in all areas—ecology, technology, social competence and political networking.

> "The crisis inside of us and the crisis in the environment are two parts of the same whole and can only be solved from that perspective."
>
> Dieter Duhm

> "Tamera is one of the ecovillages in our global network that we always listen to, as you are taking very radical and innovative steps in many directions and with great courage—such as your peace-work, university and Solar Village project. Experimenting means finding new paths and this often awakens fear and opposition. But humankind needs this kind of experimentation. We hope you will continue with your courageous initiatives so that we may all learn from your successes and failures."
>
> Ross and Hildur Jackson (Denmark), founders of the Global Ecovillage Network

Top: The house of three archways.Below: The auditorium of Tamera is the largest strawbale building of the Iberian Peninsula.

Through this combination, a worldwide network of different groups and initiatives has developed, all connected with Tamera, bringing its knowledge to their projects, or creating new projects based on similar principles.

The children of the future will be able to build on the knowledge and experience of these centres. If one day our survival depends on choosing new ways; if the breakdown of large economic and supply systems comes close; if whole landscapes become uninhabitable or social unrest threatens a peaceful way of living together, then these centres will be catalysts for a new beginning. In this way, Tamera wants to do service to the world.

Water is life

Tamera is developing a model project for natural water management and the natural regeneration of damaged landscapes. "Deserts do not develop because of a lack of rain, but because humans deal with water in the wrong way," says Bernd Müller, who is responsible for developing Tamera's water retention landscape.

Since 2007, Tamera has been building a water retention landscape inspired

Before and after creating 'Lake 1' and the water retention landscape.

by the Austrian permaculture specialist Sepp Holzer. It shows how severely damaged landscapes can be restored to their original, healthy, natural state. In place of the dusty road which was here a couple of years ago, there are now lakes, pools and retention ponds, as well as contour-following ditches (swales), 'keyline' cultivation, terraces, integrated pasture management, and continuous mixed afforestation. Around the lakes, raised beds and terraces meander with many varieties of vegetables, fruit trees and cereals.

The 1000 m² of garden area are managed in a modern crop rotation system where vegetables, cereals and green manure with mulch production take turns on each bed. We use mulch and worm compost to fertilize the soil. On 2000–3000 m² of sloping land between the terraces, we see polycultures of fruit trees and bushes flourishing. They form diverse niches and biotopes.

After only a short time, life reappears on the shores. The plants, especially the trees, are once again provided with water from below as is appropriate to their nature. The greater the variety of habitats, the greater the variety of flora and fauna. The greater the variety of fish, insects, snails and crustaceans, the more stable the ecological balance in and around the water, and the more independent the whole system becomes from fertilizers or other additional needs.

Above and below: solar education in the Solar Testfield site. Right: the Scheffler Mirror and solar cooking oven.

Above: the Solar Testfield. Top right: the solar-powered kitchen. Below: Tamera founders Sabine Lichtenfels and Dieter Duhm

Solar Testfield

Summer 2004 saw the beginning of a 'solar power village'. The solar technologies invented and developed by Jürgen Kleinwächter and his team are the core of ongoing research. Living with the technology is an essential component of the research; it is in this way that the technology is further developed and adapted to life.

The heart of the Testfield is the solar kitchen, in which a Scheffler Mirror, biogas systems, solar box-cookers, solar water heaters and a solar tunnel-drier for food preservation are in daily use. Tamera harvests about 60% of its annual electrical energy needs from its photovoltaic system. In the next years we aim to have 100% energy autonomy.

The outlines of a possible solar age become visible: a life no longer based on separation but rather on contact with nature and her cycles. This is energy autonomy in place of border security.

The silicon valley of peace or paradise under construction

There will be no peace on Earth as long as there is war in love. A culture of trust starts once lovers no longer have to lie to each other or leave each other if they turn lovingly or sexually towards another person.

Love school

An ethic is needed that enables us to become truthful, independently of how our personal life looks at any moment—whether we live in celibacy or marriage, in monogamy or polyamory. Tamera's Love School, founded by Sabine Lichtenfels, is a place for deep exchange and learning in love, sexuality, partnership and ethics. It serves the internal education of Tamera's coworkers, and the knowledge is also offered to guests in seminars.

There will be no peace on Earth as long as there is war in love. A culture of trust starts once lovers no longer have to lie to each other or leave each other if they turn lovingly or sexually towards another person.

Tamera as a peace university

Tamera is an education centre for the future. Young peace workers come from all over the world to study the knowledge and skills needed to build peace villages and autonomous settlements. The main fields of study are ecology, energy autonomy, and inner and outer peace work. The basis of these studies is learning to build community and gaining social competence.

How Can One Believe That a Local Model Can Have a Global Effect?

"Together with all life on Earth, humankind forms a holistic system. The whole acts in every detail, and vice versa: whatever happens in one part has an effect on the whole. This effect can be minimal, but increases with the significance that the local change has for the whole. In the case of a high significance, a process develops in the whole that can be described by the terms 'resonance', 'iteration' or 'morphogenetic field building'. This is the process by which peace can spread worldwide. The decisive factor for the success of such peace projects is not how big and strong they are, compared to the existing apparatuses of violence, but how comprehensive and complex they are: how many elements of life they combine and unite in themselves in a positive way. Evolutionary fields do not follow the laws of 'survival of the fittest', but rather 'the success of the more comprehensive'."

From Project Declaration 1 by Dieter Duhm, 2004

Location: Monte do Cerro, Southern Portugal
Established: 1995
Area: 134 ha (331 ac)
Population: 170
Common areas: Global Campus, Aula (strawbale auditorium with 400 seats), Terra Nova School, Solar Testfield, Guesthouse, seminar rooms, Aldeia de Luz workshops, school and kindergarten, art place, political ashram, stone circle.

Above: the Tamera school.
Left: Women power—Sarah Vollmer (left) and Sabine Lichtenfels (right) with Vassamalli Kurtaz from the Toda tribe in India.

Sieben Linden Germany
Community in diversity

by Martin Stengel

It all started with the idea to create a 'social-ecological model-settlement', born out of permacultural thinking, and with the aim to reintegrate all aspects of human living within an individual's reach. Seeing how disconnected people in modern society are from natural systems that provide food and shelter, clothing and health, we wanted to bring these systems back into people's awareness. How else would it be possible to act appropriately and creatively in response to the challenges of an over-used planet and a specialized society? How could we make this attractive, collaborative and economically viable, if not within a community of shared vision and ownership?

The idea was proposed in the exciting year of 1989, when the wall between two separate political systems came down. Enthusiasm was high; the open canvas of East Germany was looking for new concepts for its reunification with the West, and seemed to provide enormous possibilities. An association of 'Ecovillage Friends' and a cooperative were formed, and the ecovillage concept was presented all the way up to the regional politicians, that supported its implementation. However, it would still take eight years to start. Finding land, and getting permission from the planning authorities to set up a completely new village, proved to be very challenging. We bought an abandoned farm with 22 ha of land in the Altmark, East Germany. Some of the reasons we started here were a major with tears in his eyes who welcomed us as new citizens, and an old story of Poppau, being 'In the middle of the world'. Our village began without needing to borrow money to buy a lot of land, in the least populated area of Germany. An old picture of seven lime trees in front of the farmhouse inspired our name: 'Ecovillage Sieben Linden', German for Seven Lime trees.

Some of the 'pioneering settlers' lived in caravans to start populating the former monoculture of fields, forests and ruins, and worked passionately on our first communal building, which is still the communal centre. The first of August 1998 saw 800 guests for our formal inauguration, which somehow, magically, convinced the planning authorities to finally approve the development plan for the entire village on the very same day! Pine monocultures gave way to tens of thousands of newly planted trees of all varieties, fields were turned into diverse vegetable and fruit gardens, a water system with a swimming biotope was made, and new wells and a sewage system with plants were established.

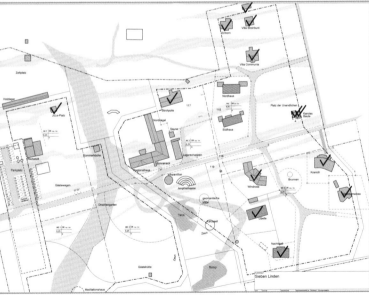

Felling a tree.

Location: Ökodorf Sieben Linden, Beetzendorf, Saxonia Anhalt, Germany
Established: 1997
Area: 100 ha (247 ac). 64 ha (158 ac) forest, 25 farmlands, 7 ha (17 ac) garden, 6 ha (15 ac) construction area
Population: 140
Common facilities: Community centre, seminar centre, bar and cinema, dance hall, workshops, forest kindergarten, farm, etc.

Fetching trees.

Hands-on building.

Meanwhile, the need for proper living space, especially for families, grew. In 2000, we started to build the first residential houses in a controversial way: modern and rationally designed post and beam construction insulated with recycled cellulose fibre, and 'Villa Strohbund', built only from locally sourced natural or recycled materials and without machines.

Self-responsibility

Sieben Linden ecovillage is a social and ecological settlement based on the principles of self-sufficiency and self-determination, and designed to accommodate approximately 300 people. We all strive for manageable and transparent social structures that promote self-responsibility and help to strengthen cooperative relationships between human beings and nature. 'Community in diversity' is one of our main slogans, within which there are different ideas for a sustainable lifestyle. A diversity of more or less spiritual worldviews, ways of making a living, levels of commitment to serve the community, and strategies to connect to the world around the village, continuously contribute to forming our 'community'. Originally, we had intended to provide room for this diversity within the pattern of neighbourhoods as subcommunities.

These are caring strongholds of mutual social support inwards, while forming a colourful mandala of interconnecting diverse lifestyles and cultural contributions to the village community. We divided the village into several building zones, with multiple housing units, mostly with communal kitchens over-arching family structures. Sub-communities like 'Brunnenwiese', 'Club99' and 'Globolo' (inspired by Hans Widmer's bolo'bolo, an autonomous community in a utopian ecological future) formulated their own internal interpretations and cultivate their expression of ecovillage life in the four dimensions of social, spiritual, ecological and economic life. The most radical and famous became 'Club99' with its well-known characteristics: a fully shared economy, a vegan diet free of industrial packaging, construction sites without power tools and new industrial materials, an intense practice of communal living, including open relationships (non-obligatory certainly!). This is where 'Villa Strohbund' was built and the European directory of ecovillages Eurotopia was published. Working horses went into the village forests and the cooperative's fields, and a wild-herb and raw-food business started, which later became Raw-Living Germany.

Strohpolis.

The following decade taught us a difficult lesson. The pioneering spirit, which had motivated ecovillagers to give their utmost to communal needs, while limiting the individuals' need for private space and time, gradually weakened. One of the reasons was the large number of social spheres to be cared for, organized and communicated with. Neighbourhood members had their days and late evenings filled with meetings in three circles: the village community, the neighbourhoods, and last but hopefully not least, their own family— not to mention their intimate relations with a partner. Parents were the first to give up, but others developed a need for more private space and time. To me this is one of the most interesting questions; how can a community combine healthy individual expression and fulfilment, with a continuously inspiring and rewarding, but also demanding community life?

A high ecological standard

The village has become an open and hospitable settlement with an ever-growing guest and seminar business.

The complex self-organizational framework unites specialized working groups and six councils under the umbrella of a general assembly consisting of all cooperative members. Here we apply modern sociocratic principles. The community organizes three 'intensive periods' every year, during which questions of mutual support, further development of the vision, fundamental economic issues and many more are discussed. Cultural offers include regular dance classes for all ages, yoga lessons, community gaming nights, bar and cinema. Around ten young people have a socio-ecological year as volunteers with us, and our wonderful forest kindergarten welcomes villagers and children from the larger area.

The village has an exceptionally high ecological standard, which resulted in the lowest ecological footprint of all ecovillages in Europe (study by the

Strawbale building.

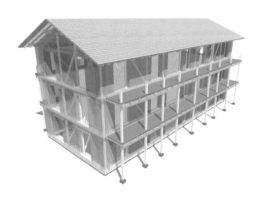

University of Kassel, 2006). The main diet is vegetarian or vegan food grown in our own gardens. Wastewater is cleansed in reed beds and re-used as irrigation water for 2 ha of vegetable gardens and fruit orchards. Dry compost toilets safely compost human excrements. Heating systems use wood from our own forests and large thermal solar panels. Photovoltaic panels of 50 kWp provide electricity for the village-owned grid. The entire village is free of Wi-Fi and mobile phones. The area covered by buildings is limited to 16 m² (172 ft²) per person to keep the sealing of soils as low as possible. All land property, as well as the main shared buildings, infrastructure and most of the dwelling houses are owned by our two cooperatives, so every villager is a co-owner of the whole.

'I take responsibility!' is one of our seven commitments. I am looking forward to the flow of new responses that we will develop to our internal and global questions!

Villa Strohbund

Villa Strohbund is a two-storey house of 100 m² (1076 ft²) floor space, designed and used as the communal living space for 'Club 99'. It has a kitchen and dining room, a big room for meetings

Villa Strohbund.

Top: Libelle. Bottom: Brunnenwiese.

and gatherings, a naturally cooled pantry and a storeroom. It is the first legally approved strawbale building in Germany, and this became one of the milestones within the rapidly growing strawbale movement. In 2003 the first two strawbale domes in Europe were built, one as a communal bath, the other as a guestroom. Since then all residential buildings in Sieben Linden use an ever-evolving variety of strawbale technologies that contribute to the scientific research and practical development of strawbale construction in Europe.

The foundation of Villa Strohbund is made of recycled granite blocks, using a total of only 40 kg of cement. The timber frame is made of round timber, cut in our forests by hand, hauled by horses and assembled by hand with the use of traditional tools. Walls, floors and roofs are straw bales from our organic harvest. Clay and sand come from Sieben Linden. The building cost less than 15% of the average in Germany, but it took around 15,000 hours of work by professionals and many students. An ecological footprint

analysis undertaken by the Technical University of Berlin, shows that the whole construction process of Villa Strohbund emitted less than 3% of the CO_2 of an average modern residential house in Germany.

Brunnenwiese

Brunnenwiese is a neighbourhood of villagers working in the field of human health with both conventional and alternative approaches. Its Spiral House is the first of three houses, designed and built in 2005-06 by the members of the neighbourhood. A large kitchen and living room, a bathroom and a separating toilet provide living space for about ten people. The house has an organic spiral shape, with a big tree trunk at its centre. The stairs add a vertical dimension to the dynamic of the spiral turning around the central tree.

A large glazed room on the second floor, used for social and spiritual meetings of the house community, crowns the idiosyncratic design. The Spiral House is a strawbale house, with a 'warm core',

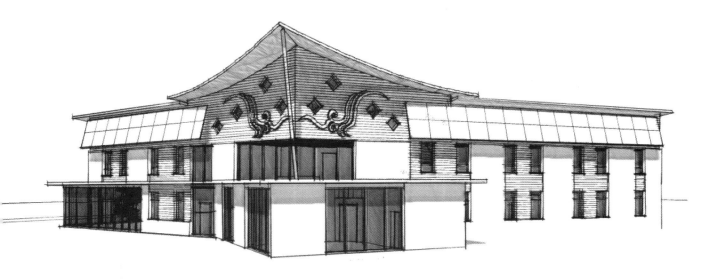

New guesthouse.

a small room in the centre of the ground floor, containing the central furnace. This room naturally spreads its warmth by radiation and convection to all rooms on all levels, and is used as a bio-sauna for drying such things as clothes, herbs, fruit.

Libelle

Libelle, meaning dragonfly in English, was designed in 2009 to further reduce wood for heating. In our experience, even, the best insulated houses still use a lot of wood, so German forests cannot supply enough. Housing ten people on 312 m^2 (3360 ft^2) in several apartments, Libelle has a central large hot water storage of 13.5 m^3 (477 ft^3). This is heated by 63 m^2 (680 ft^2) of thermal solar panels and a wood fuelled boiler. The walls and roof are insulated with straw bales and all windows are triple-glazed, including the fully glazed southern façade. A ventilation system with heat recovery further reduces heat losses. This concept has reduced the amount of wood fuel needed to only about four wheelbarrows full per person per year or below 30 kWh/m^2/year (3 kWh/ft^2/year).

Early 2018 will see the start of our most ambitious building project ever. For more than ten years, we have been working on designing and financing a central guesthouse for up to 40 people, with large seminar rooms, dining room, offices, lounge and café. Now we are ready to build 1,200 m^2 (12900 ft^2) of floor space with a budget of over 2 million euro. The building will frame the south and west of our central village square, which has been waiting to be completed and filled with activities since the inauguration of the village. This building will move the focus from our private community yard to a more public space, while increasing the capacity and standard of our seminar business, which is the core economic activity at Sieben Linden. The building will be built according to Sieben Linden's high ecological standards, with straw bales, wood and lime, separating toilets and more than 100 m^2 (1,075 ft^2)of thermal solar panels.

Chololo Tanzania

A model of good practice

by Michael Farrrelly

Tanzania is highly vulnerable to the impacts of climate change. Eighty percent of Tanzanians rely on climate sensitive rain-fed agriculture for their livelihood. Reducing vulnerability to climate change is crucial for strengthening socioeconomic development and assuring food security.

Funded by the EU's Global Climate Change Alliance, the Chololo Ecovillage project supported the community to test, evaluate and take up 25 innovations in agriculture, livestock, water, energy and natural resource management.

The project was delivered by a partnership led by Institute of Rural Development Planning, with Hombolo Agriculture Research Institute, Dodoma Municipal Council, and Tanzania Organic Agriculture Movement.

A gift of chickens

The huge hawk circled high over the baobab and thorn-bush landscape of Chololo Ecovillage, its white wing flashes clearly identifying it as a Bateleur eagle. Gazing up in awe at the sight of this monarch of the skies I asked Chololo farmer Mary Mpilimi, "What do you call that bird in Swahili?" She glanced up and snorted with indignation, "Mwizi wa kuku – that's the bird that steals my chickens!"

Life can be harsh in Chololo, a drought-prone village in the semi-arid centre of Tanzania. Chololo has experienced six major droughts in the last 30 years, and has been food insecure for much of that time. With climate change, temperatures are increasing; rains are reducing (3.3% per decade) and becoming less predictable, leading to crop failure, poverty and hunger. Women walk five hours to fetch firewood from the much depleted forest, and in the past, trekked to neighbouring villages in search of drinking water. Food aid handouts and seasonal migration in search of casual labour have been the norm for many years.

Mary is one of the dynamic early adopters, who jumped at the opportunity to try out some of the 25 climate change adaptations introduced by the EU-funded Chololo Ecovillage project. Designed to build communities' capacity to adapt to climate change and address the problems of drought, hunger and poverty, Chololo Ecovillage is now recognized nationally as a model of good practice, working across agriculture, livestock, water, energy and forestry, changing the lives of the 5,500 residents of this extremely vulnerable community. Mary explains: "I have learnt a lot about adapting to climate change. In the past

Mary explains the benefits of the Chololo Ecovillage project. Right: the fish farming project.

we were planting traditionally, but now we have learned to plant in measured rows, and we are getting higher yields. I used to keep my chickens outside and the dogs and hawks took them, but now we keep them safe in a chicken house. I used to beg money from my husband for things like clothing, medicine and school fees, but now I don't; I can just catch a chicken, sell it and get money. Right now my children are going to school and I can buy materials for building my new house."

In the past, with traditional farming methods, the frequent droughts led to poor yields. Now, with training, Chololo farmers are following good agricultural practices, planting at the right time using improved drought-resistant seeds and farmyard manure.

Mary says, "I advise other women like me to try and plant even on a small portion of their land—to see the difference you get from planting in rows with correct spacing. If you do that in two acres you can get as much yield as three or four acres before."

Crossbreeding local hens with exotic cockerels combines the adaptive attributes of the indigenous chickens with the high producing abilities of the exotic chicken. Mary was one of 120 chicken keepers given a rooster and trained on local chicken management including feeding, rearing, breeding, record keeping, housing and disease control. "I was able to sell 100 crossbred chickens for one million shillings (500 Euros). If you have a problem, like the children are sick and need to go to hospital and your husband is not around, you can just take two or three chickens, sell them and fix the problem."

New and old solutions

Chololo Village Chairman Michael Mbumi summed up: "In my village everybody

Improved goat breeds and local production of leather goods diversify livelihoods.

has benefited from the project. Nobody is going out of the village in search of food. Those who have shortage get food within the village from farmers who have enough to spare."

The project has introduced a package of ecological agriculture technologies to make the most of the limited rainfall, improve soil fertility, reduce farmers' workload and improve the quality of local seeds. These include:

- Ox-drawn tillage implements are improving farming methods.
- Crop residues are being used to feed livestock.
- The introduction of improved breeds of cattle, goats and chickens increased the productivity of the animals, producing more meat and eggs more quickly.
- Soil water conservation measures help to capture rainwater and prevent soil erosion.
- Soil fertility is improved through use of farmyard manure.
- Improved early-maturing, high-yielding seed varieties have rejuvenated the village seed system. All the seeds are 'open pollinated'(not hybrids) so they can be recycled by farmers year on year. Community seed production ensures that a good supply of quality seeds is available for planting each year.
- Good land preparation improves soil and water conservation. Optimal plant population with correct spacing distance, intercropping and crop rotation improve yields per acre and help control weeds, pests and diseases.
- Leather tanning diversifies livelihoods. The project trained people in vegetable leather tanning using Mimosa tree bark extracts. This increases the value of each goatskin tenfold.
- Production of leather goods such as shoes, belts, key holders and phone covers can fetch more money than selling leather. Diversifying livelihoods makes people more resilient to climate change.
- Modern beekeeping trebles honey production. Four beehives in an acre of sunflowers can increase crop yield by 30% through improved pollination.

- Some farmers built and stocked fishponds, harvesting adult fish for household consumption, and selling fingerlings to other fish farmers. Once harvested, the fish keepers are using the water in their fishponds to irrigate trees. Fish is a rich source of protein for their families.
- School roof rainwater harvesting provides 60,000 litres of fresh water. A hand pump lifts the water for drinking, washing clothes and watering the school tree nursery.
- Solar powered village water supply is cheaper to run and more reliable.
- Subsurface and sand dams capture thousands of tons of rainwater for livestock watering.
- Tree planting and agroforestry has been increased with the planting of 33,650 tree seedlings and 3,000 trees in the village forest reserve.
- Community land use planning is reducing degradation of natural resources.
- Renewable energy is reducing deforestation. Domestic biogas digesters reduce fuel-wood use to zero and cattle dung is converted into fuel for cooking stoves and lamps. Two or three cows generate enough gas to meet a family's daily

cooking and lighting needs. The fertilizer closes the nutrient cycle, and reduces soil degradation and erosion. The biogas process is carbon neutral, contributing to the reduction of greenhouse gas emissions.

The project is attracting attention from far and wide. Dr Julius Ningu, Tanzania's Director of Environment, Vice President's Office, said: "Through a holistic approach the project is breaking new ground, achieving synergies and strengthening the knowledge base of good practice in climate change adaptation, while reducing carbon footprints. The National Climate Change Strategy encourages such initiatives to build the critical mass of expertise to address adaptation challenges, while safeguarding precious natural resources and strengthening the country's voice in the global climate change debate."

A biogas digester and farmer Minza Chiwanga shows intercropping.

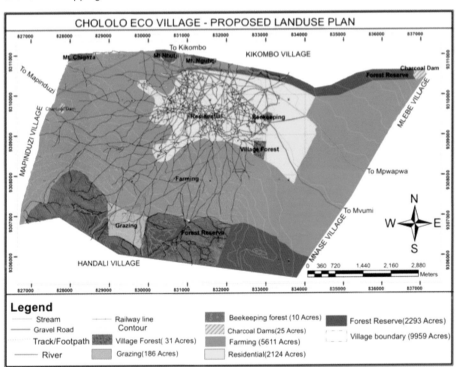

CHOLOLO ECO VILLAGE - PROPOSED LANDUSE PLAN

Legend

⎯ Stream	⎯ Railway line
⎯ Gravel Road	Contour
···· Track/Footpath	Village Forest(31 Acres)
⎯ River	Grazing(186 Acres)

Beekeeping forest (10 Acres)
Charcoal Dams(25 Acres)
Farming (5611 Acres)
Residential(2124 Acres)

Forest Reserve(2293 Acres)
Village boundary (9959 Acres)

Location: Chololo, Dodoma, Tanzania.
Established: 2011.
Area: 4,000 ha (9,900 ac) semi-arid, drought prone.
Population: 5,500.
Housing: Traditional: wood and mud construction; modern: concrete blocks and corrugated iron roof.
Communal buildings: Primary school, dispensary, ecovillage centre.

Narara Australia
Inspired by life

by John Talbott

Narara is a recent ecovillage, still in the birthing process after nearly five years of planning and consent processes. Planned for 400-500 people, the first 60 dwellings are about to begin.

Our vision is an environmentally, socially and economically sustainable world.

Our mission is to create a sustainable ecovillage as a demonstration of this vision.

Our aim is to research, design and build a stylish, inter-generational, friendly demonstration ecovillage at Narara, blending the principles of ecological and social sustainability, good health, business, caring and other options that may evolve for our wellbeing.

Narara Ecovillage is pretty normal in many ways. We're looking for:
- a village style development where we can live work and play together
- a strong community focus, building our social infrastructure as well as our physical infrastructure and buildings
- reducing our ecological footprint by sharing facilities like tools, meeting rooms, offices, laundry, which avoids the need to duplicate in every house

Sunset in Narara.

- planning activities together like community gardens—a great way to be together, get healthy food and have fun
- incorporating commerce on site, reducing the commute for many of the residents to a walk or bike ride
- developing education and visitor programmes to bring people to Narara to experience sustainable living first hand.

Narara is also unusual in some ways:
- We're a co-operative where we are the funders, developers, clients, decision makers and future residents.
- We aim to be mixed use, so we're not just a residential community, but also a thriving village with businesses and cottage industries, agriculture and food production next to conservation and wildlife preservation. Town planners hate this but it's how traditional villages have always been!
- We have a 'Community Title' where the members own and maintain all the infrastructure (not the local council).
- We have community owned utilities for water and electricity supply.
- We use Permaculture—a system of agricultural and social design principles that uses patterns found in natural ecosystems to inform our

Above and left: Aerial views and a viewof the valley.

human design process, with care of the Earth, care of people and fair share as core tenets.

- We will develop a green transport scheme with priority given to pedestrians and bicycles on site, and reduce car use through car sharing, community mini bus and use of public transport.
- We use Sociocracy, an inclusive governance and decision making process that has helped us in developing our community structure, bylaws and common agreements.
- We aim to be carbon neutral—and maybe even carbon positive—where we eliminate CO_2 emissions through using 100% solar generated energy in buildings and transport with Smart Grid technology.

Not just for the eccentric greenies, but the way of the future

As our local council will tell you, we don't tick many (or any) of the normal boxes for residential subdivisions and it makes them a little crazy, which made it challenging for us. However, what is great about Narara is that we're very much a 'middle class ecovillage'—not too weird, sort of normal looking for the most part, and something that the general population can relate to. Having lived in a remote ecovillage for many years that was definitely on the fringe, I have been so impressed with how this diverse group of strangers has come together in

Heleen and the children.

a short space of time to develop a real life working model for sustainable living. The fact that it is easily accessible to a large metropolitan population of five million, just over an hour away by public transport (or car), only increases the potential positive impact Narara will have. We hope we can show that ecovillage living is not just for the eccentric greenies, but the way of the future.

Ecovillage principles

- The landscape dominates; housing sits in a natural setting in native forests. Views to nature are key—the valley and the common gardens.
- Common gardens between houses provide pathways, shared spaces for social interaction, food production, recreation and contemplation.
- Permaculture Principles employed in the design of all gardens.
- Streets as social spaces: shared zone for pedestrians and vehicles with max speed of 10 km; street trees to provide shading, amenity, social space. Cars and parking kept to the periphery with pedestrian and bike priority.

- Soft landscaping elements, shrubs and trees demarcate private from public spaces rather than fences.
- House lots are orientated to maximize sun access—each dwelling to provide enough solar PV to cover annual energy consumption.
- Integrated water cycle management, wastewater treatment and reuse for toilet flushing and gardens; all drinking water from on-site farm dam; stormwater collection, retention and reuse.
- Smart grid: pioneering an intelligent community scale electricity grid for the village, including electricity generation, battery storage and peak demand management.
- Smaller houses encouraged with high energy efficiency and use of natural and sustainable materials; community has developed their own building standards.
- Encourage on-site businesses to provide employment for members and economic sustainability for the community.

Location: Narara, New South Wales, Australia, an hour north of Sydney.

Established: 2012

Area: 64 ha (158 ac), 35 ha (86 ac) biologically diverse native forest, 15-20 ha (37-49 ac) for development, the rest for food production. Over 120 species of birds identified.

Population: 450-550 when fully developed in 7-10 years. First development of 60 dwellings currently underway. Current membership of 160, age range 0-85, average age 45-50, including 35+ children.

Housing: 150-180 dwellings with diversity between single-family dwellings, terraced town houses, apartments and shared houses.

Communal buildings: Office building and admin centre, Visitors Centre, meeting hall, amenities, workshops and educational facilities, commercial space for rent and agricultural land. A 'cohousing' style common house for each stage.

Site Plan Stage 1
NARARA ECOVILLAGE
1.16

Left: Birdwatching and bushwalking. Below: Approaching Narara in autumn, and indigenous residents.

Hurdal Norway

Making sustainable life accessible for the mainstream

by Simen Torp

The story of Hurdal Ecovillage

After more than ten years of planning, the first new Norwegian ecovillage gains momentum; linking up with the business world, politicians and other mainstream players, it is the hub of a growing 'Sustainable Valley'.

Establishing Norway's first new ecovillage has been a long, demanding and exciting journey. Our idea was to buy a farm and turn it into an ecovillage. In 1998 we were more than 100 people in Kilden ('the source') ecovillage group, trying to find the 'right' place. We visited more than 40 potential sites in three years. Finally, the mayor of Hurdal municipality contacted us; they owned a parish farm, called Gjøding gård. I had a strong sense that this was the place. The farm had 16 ha of fertile farmland, a lot of forest, a beautiful beach by the lake and a river nearby. Hurdal municipality is 30 minutes north of the Oslo airport Gardermoen, with 2,700 inhabitants (in 2016).

However, most of Kilden's members wanted to live closer to the city of Oslo. Only my wife Kristin and I, and another young couple decided to focus on Hurdal. In 2002 we were there: two young families with a vision of an ecovillage. I was 24 years old, with no formal education, no money, no experience in building or creating a community. But I had two important things—a clear 'inner voice' and a limitless, powerful vision that was slowly turning into a life mission.

After meeting with the administration of Hurdal municipality we realized that we had to create a master plan for the whole area, to be able to build the ecovillage. The Gaia architect Rolf Jacobsen agreed to help, and the process of turning the farm into an ecovillage began. With a group of ten adults, we moved to Hurdal. We made an agreement with the council to rent the farm, while we were working on the plan.

As the Gjøding farm and church have historical significance, we had to consult archaeologists. It was a shock to hear that the senior archaeologist would never accept us building new houses there. The green pine forest just behind the ancient church was protected. We tried to argue that our plan was to create a green, permaculture fruit forest, but that didn't help. The project seemed impossible to continue, since nothing could be built. One person had stopped the whole project. Then something magical happened: a freak storm blew down the whole pine forest in the exact area where we wanted to build. This effectively

Active houses in Hurdal ecovillage.

ACTIVE HOUSE NATURAL BUILDING:
- Efficient space use and flexibility
- Good quality and durable
- Easy to use
- Built with environmentally sound materials
- Energy efficient
- Uses renewable energy sources
- Natural hybrid ventilation
- Comfortable with a good indoor climate
- Rich in outdoor areas
- Well designed

removed the reason for stopping the project, and silenced the archaeologist. We were able to continue our process.

In 2004 we got a loan from the only ethical bank in Norway, Cultura, to buy the farm, a horse and a tractor. After three years of hard work to make the new master plan, the local council gave us approval. In 2003–04 we built eight small temporary houses close to the farmyard as places to live while working on the bigger building project behind the farm. At first, our building philosophy was to self-build natural houses. However, we had no experience, many small children and little money—a very difficult start for our group. Most people building their homes were exhausted before their house was finished, so we had to abandon this concept.

Aktivhus—Active House

We tried to find a building company that could help us build small, simple, natural houses: 100% wood, wood fibre insulation, no plastic in the walls, self-sufficient in energy and electricity, a wood stove, non-toxic materials and natural ventilation. But no companies were interested. After two years of searching, we decided to start our own building company called 'Aktivhus'. After five years in the market, and to our surprise, the company is a success. Every week people contact us interested in buying or building an Aktivhus.

In 2011, it became clear that our village group was not able to finance and build what we had planned, most people were tired and we were close to bankruptcy. At this point, I made contact with an experienced building entrepreneur, Pål Lund-Roland. I had a strong feeling that he could be the man to help us. It is the first time that I have met a businessperson who understood our holistic lifestyle and valued our project and its philosophy. Together with Pål and

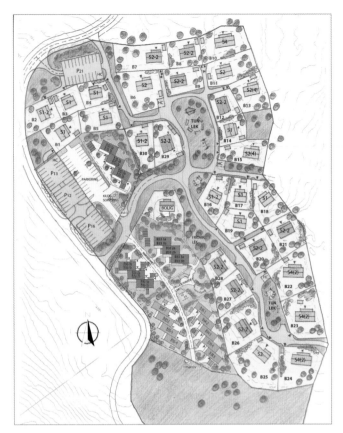

Above: An Active House.
Left: Master plan of Hurdal Ecovillage.

his team, we developed our concept. They were responsible for the 'hardware', while we, the ecovillage pioneers, took responsibility for the content of community life. Pål created a company to develop the Active House concept further for industrial production. Because of the high start-up costs with the infrastructure and other innovative solutions, there had to be 70 houses and apartments built in the first construction phase, a lot more than in our original plan. The total price of all the 'hardware' in the first building site was around 2 million euros.

Until then, the way to join the project was to attend an introductory course, apply to be a member and wait for approval. Now I had the task of selling the houses on an open market. An Aktivhus apartment with a plot costs between 200,000 to 540,000 euros. In August 2013, when we had sold 22 houses, the developers

started work and in a few weeks the new village road was ready and the first foundations were laid.

We have now started an ecovillage company called Filago, that is a driving force for making sustainable homes and an eco-friendly lifestyle available for everyone. Filago is founded on the triple bottom line of sustainability, the three P's: People, Planet and Prosperity.

In the autumn of 2013, neighbours, visitors and the media came to see the construction site. Finally something 'big' was happening in little Hurdal. Even conservative locals started to become proud. Other municipalities contacted the Hurdal council saying: "You guys are lucky to have this fantastic ecovillage project." In November 2016 more than 160 people were living in Hurdal Ecovillage and 70 houses had been sold.

Above: The café and bakery are already a popular meeting place for both locals and ecovillagers. Left: Interiors of Active Houses.

We plan to be fully developed by 2020 when the ecovillage will consist of 200 apartments or houses with about 500 people.

Since starting our project we have aimed to build bridges to authorities, politicians, neighbours, entrepreneurs, banks, media and the rest of mainstream society. I don't believe in creating isolated islands, but rather an attractive 'lighthouse' that demonstrates sustainable practices and consciousness, and attracts positive attention from all levels of society. The time has come to demonstrate 'luxurious simplicity'. People will understand that everything we do in ecovillages is based on common sense. You can experience, touch and understand sustainable solutions. Most people want good food without pesticides, to live in a healthy house, not to destroy our planet, more peace and love in their lives, less stress and a more healthy life, and being able to spend more time with their family and neighbours.

To me, the ecovillage concept is all about creating win-win-win solutions. One of the key factors is to build bridges into the financial world. Many wealthy people are ready to invest in what we are doing. Deep within, we are all 'farmers' who want to sow our seeds in fertile soil, take care of the small plants that come up and share and enjoy the harvest together with people we love and care about.

Framtidssmia: Smithy of the Future

Close to the ecovillage, we have bought an old local school, with 2,200m². A Visitors Centre is being developed, with an eco-playground, space for offices, workshops, shops, a wood-fired bakery with Schauberger mills, a village café, and a holistic health centre. We want it to be a 'sustainable experience centre', where people and families can come and taste delicious organically produced food from our own fields and be taken on a tour through the ecovillage. The café and bakery are already a popular meeting place for both locals and ecovillagers.

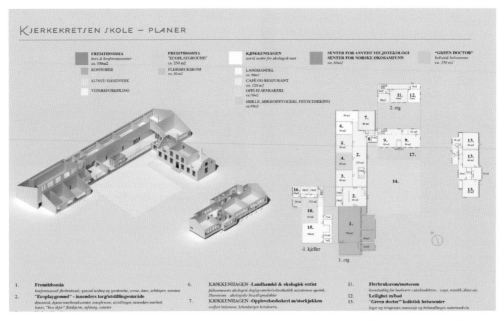

The Smithy of the Future Visitors Centre is being developed, with space for offices, workshops, shops, a wood-fired bakery, a village café and a holistic health centre.

Above: Lokal bakery, and a summer concert.

Established: 2013
Location: Hurdal, 1 hour north of Oslo, Norway.
Area: 60 ha (150 ac) farmland.
Homes: Single houses, detached housing, apartments.
People: 2015: 150 people.
Common areas: Framtidssmia: local community centre, parish house on farm, bakery, health centre, Gjøding Farm.

Hurdal 'Sustainable Valley'

by Frederica Miller

The ecovillage in Hurdal came about thanks to a visionary municipality. As the ecovillage has been built the municipality has also developed its visions and plans for the future. The concept of Hurdal becoming the first Sustainable Valley was born. Close to the airport and Oslo, with the first new ecovillage under construction, Hurdal can be Norway´s chance to take the lead into a sustainable, carbon neutral future. We don´t need more Silicon Valleys, we need Sustainable Valleys. Runar Bålsrud, the mayor of Hurdal, has endorsed this vision, and in 2014, a unanimous municipal council decided to make Hurdal a Sustainable Valley! This is also an attempt to convince the Norwegian government to set aside 1% of its oil income for sustainable development.

The Sustainable Valley project has started with a municipal plan for a new local centre—a sustainable urban village—planned to house 3,000 new inhabitants, new green businesses and a Sustainable Academy. Frederica Miller from Gaia-Oslo architects is project leader in a team with architects Helen and Hard. The plan is being used to simultaneously encourage and develop local new green businesses. So far, several projects are under way:

- A wood project connecting local forest owners, a sawmill, and the Active House concept, with the goal of producing building elements and carbon neutral houses made entirely out of local wood.
- A sewage project—to utilize the nutrients in sewage for forest or food production.
- A food project—to focus on local organic food production, connecting with restaurants in Oslo.
- A Green Business project with the local business group to look at certification practices for existing and new local businesses. In addition, Hurdal municipality is reviewing its internal services.
- A transport project to establish a local car sharing pool, and improve public transport in Hurdal.

Tasman Ecovillage Australia
A village within a village

by Karen Weldrick and Jane Morton

Tucked away on the magnificent Tasman Peninsula, our community is Tasmania's first ecovillage. Overlooking Parsons Bay, surrounded by farmland and native forest, we sit on the outskirts of Nubeena—a small seaside, holiday and fishing township. Nocturnal wildlife abounds. We wake to a dawn chorus of native birds, and wallabies descend from the hills to graze around the village in the morning and afternoon light. We think of ourselves as a village within a village—the Nubeena shops, a medical centre, school, local produce stores, a community bank and a library are all within easy walking distance of our community.

The region is home to an extensive range of stunning bush walks, first-class surf breaks and sheltered bays and coves, ideal for swimming, deep-sea diving and fishing. Tourism is this area's life-blood, and we are just a ten-minute drive from the highest sea-cliffs in the Southern Hemisphere, the internationally iconic Three Capes hiking track, and Tasmania's premier tourist attraction—the historic Port Arthur convict settlement.

Getting started

Our founder, Ilan Arnon, was born in Israel and lived in a kibbutz in his early years. He

Greenhouse construction.

had a life-long dream of establishing a sustainable intentional community, and in 2007 he purchased a run-down 1970s motel and golf course with that end in mind. There were some major obstacles to overcome, including putting in place the right legal and financial structures, gaining council approvals and surviving the global financial crisis. In 2010, the council approved the current village design.

In 2011 several forums on 'Intentional Communities—an answer to a global chaos' were held around Tasmania and these attracted the people who became our founding members. Later that year, over a long weekend of intense and focused dialogue, we agreed on our core values and vision. Soon after that, we formed the Tasman Ecovillage Association, known as TEVA, our not-for-profit incorporated members' association.

Over the next 18 months, the founding members and several specialist consultants refined and confirmed the village masterplan and completed the registration necessary to become a Community Development Scheme (a private title with common ground, with membership requirement for purchase). In June 2013 Tasman Ecovillage was officially born.

A demonstration project for sustainable living

In 2013, some of the founding members purchased apartments that had been part of the original motel, and these provided immediate accommodation while the rest of the village was coming into being. Next, we constructed a ring road with five residential clusters ('pods'), surrounding the central hub and installed underground services for the first ten residential home sites ('seeds'). In mid-2015, we released these plots for sale. Seven sold within a year and three houses were completed in 2016. The next phase of providing infrastructure to the remaining clusters is already underway, enabling titles for most of the remaining plots to be released for sale soon.

In early 2017, we have 24 permanent residents with ages ranging from 2 to 75, as well as temporary residents. Some people own their home while others are renting or camping. We are a diverse group who integrate, to varying degrees,

family, work and community life. We each contribute in different ways.

It's an exciting time with new homes being planned, delivered and built. The cost of land is very affordable by Australian standards, making the project attractive to people on lower incomes and those downsizing from big city homes. We are already seeing resourceful solutions to housing ourselves without big mortgages—these include tiny houses, a straw bale house and a prefabricated home which recently arrived on a truck. We recycle building materials where possible, and are exploring innovative ways of repurposing 'dongas' (transportable workers' accommodation) for permanent housing.
The Village Hub Bar and Café and Village Visitor Accommodation are central to village life and bring many benefits including employment and enterprise opportunities as well as visitor accommodation. The nine two-bedroom apartments within the original

motel are the mainstay of the residents' accommodation and several have been recently renovated.

'The Shed', originally a machinery storage area, is now our community kitchen/ dining area and common room. Here, residents and visitors meet, eat, play music, watch movies, conduct workshops, cater or just hang out. Our weekly potluck community dinner is held here. We converted an adjoining building for workshops, yoga and general gatherings and have upgraded the spa and sauna, which were part of the motel. Having our accommodation and community facilities in close proximity create a lively village centre with a strong sense of connection. Our community gardens are our most popular meeting place. They provide us with wholesome, organically grown fruit and vegetables. Some land is leased to community members for commercial food production and some residents keep chickens and sell or barter the eggs.

Next steps

We are moving forward with our vision of 'a thriving, caring community that celebrates our connection with the Earth and each other and cultivates a sustainable, peaceful and productive lifestyle'. We are learning from other intentional communities and reach out to experts for assistance.

We are gradually developing robust and resilient approaches to decision making and conflict management and have adopted the sociocratic method as part of this process. We are exploring how best to balance the relationships between Ilan as the developer and operator of the commercial enterprises, our Body Corporate (the property owners in the village), and our members' association,

TEVA, which includes owners, renters and supporters.

Recently we entered into discussions with the indigenous Pydarerme people. We are committed to collaborating with them to ensure the respectful and appropriate evolution of the site and to embracing their wisdom, knowledge and cultural perspectives in an authentic way.

Our ecovillage is expanding as members buy neighbouring properties. A recent purchase of the 6.6 ha hillside on our northern boundary, offers us water security, bushfire protection, additional land for renewable energy, and more space for food production. Another linked property gives us private walking access to Parsons Bay waterfront for kayaking, fishing and nature walks. Employment opportunities are limited in the region, so we support our members to develop their own ethical and sustainable businesses.

We want to walk our talk, and aspire to become a demonstration project for a diverse range of sustainable living systems including power, water, food production, education and eco-tourism. It is early days yet.

As more members join us, they bring new ideas and fresh energy. Together we are moving towards something truly unique, inspiring and nurturing.

Our vision:
A thriving, caring community that celebrates our connection with the Earth and each other and cultivates a sustainable, peaceful and productive lifestyle

Location: Nubeena, Tasman Peninsula, Tasmania, Australia.

Established: June 2013.

Area: 9.3 ha (23 ac) with established organic gardens, some fruit and nut trees.

Population: 24 and growing.

Housing: Cohousing structure, up to 55 homes and nine apartments. Some owner-occupied; short or long-term rentals and visitor accommodation.

Common facilities: Community kitchen/dining area, large common room, large gathering space (converted pool area), spa and sauna, small gym, laundry. Privately run businesses.

Above: White Beach. Below: The Grotto.

The founding group. Right: Helpers in the garden.

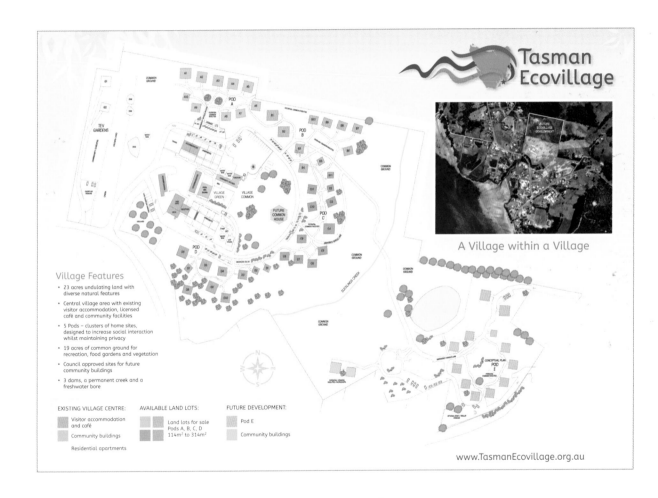

Tasman Ecovillage

A Village within a Village

Village Features

- 23 acres undulating land with diverse natural features

- Central village area with existing visitor accommodation, licensed café and community facilities

- 5 Pods – clusters of home sites, designed to increase social interaction whilst maintaining privacy

- 19 acres of common ground for recreation, food gardens and vegetation

- Council approved sites for future community buildings

- 3 dams, a permanent creek and a freshwater bore

EXISTING VILLAGE CENTRE:

Visitor accommodation and café

Community buildings

Residential apartments

AVAILABLE LAND LOTS:

Land lots for sale Pods A, B, C, D
114m² to 314m²

FUTURE DEVELOPMENT:

Pod E

Community buildings

www.TasmanEcovillage.org.au

Hua Tao China
Innovation for Chinese civilization

by Woody Cui

A group of environmentally aware Chinese entrepreneurs are taking the bold step of creating an eco-community with people who share their ambitions. Crowdfunding and sustainability will help the pursuit of a healthy and simple way of living. We actively engage in the promotion of traditional culture, and at the same time are open to the outside world by visiting other ecovillages and joining the Global Ecovillages Network.

Our key values are openness, harmony, tolerance and fair share. Our vision is to make the Hua Tao community into a Chinese laboratory of sustainability, dispelling and reducing the terrible smog. We want to show the world that life can be different. Hua Tao applies sustainable and sophisticated living systems to its resource management, which include regional biodiversity, green planning, water retention, energy conservation, refuse treatment and more. All these measures make us an ecological community with a better environment both inside and outside our community. Hua Tao has now become a pioneer in promoting the restoration of the environment, morality and the economy.

The construction of Hua Tao ecovillage will be completed in three phases. In the first phase, two residential buildings and a small public centre have been built and were opened in May 2016. We have planned for seven residential buildings in the future. We adopted the traditional Chinese courtyard style of western Sichuan. In the residential building, each family has a separate apartment with shared open space for everyone (for example: lobby, atrium, and corridor). Each residential building has a public kitchen. The first public centre outside the residential area has been established for communication, entertainment and training. A café will also be established soon.

Hua means China, Chinese tradition and culture. Tao is the way, which represents nature and innovation. Hua Tao is the pathway of tradition and innovation for Chinese civilization.

Features of Hua Tao ecovillage

- The master plan adapts to the local environment by keeping existing rivers and bamboo vegetation.
- People learn how to share and cooperate with each other in a harmonious ecology.
- An education centre for Chinese culture, Hua Tao also focuses on education reform. Compared with a traditional education model, we think more about artistic and spiritual practice, we also hold workshops on ecological training, leadership development and spiritual retreats in different forms. With our activities and unique facilities, both residents and visitors will find an approach that fosters inner peace and profound personal change.
- We are building with materials that require 90% less energy than traditional buildings. With special materials and a passive energy saving design, the room temperature will always be nice even without a heater or air conditioner.
- Building materials are locally sourced, totally eco-friendly and recyclable, with a minimal impact on the environment and human health.
- Buildings have the traditional courtyard style of western Sichuan, an open space for sharing and communication, connected with a horizontal and vertical social space, lobby, atrium and corridor.
- Hua Tao Grand Hall is a multi-functional centre for international forums.
- We are making the transition to organic agriculture, achieving a sustainable, ecological and circulatory system of agriculture, which is pollution-free, productive and easily operated. This promotes long-term sustainability. The output of organic agriculture is not just healthy food, but also better soil and balance in nature.
- Better soil conditions will help to retain water and nutrients, reducing the risk of groundwater pollution.
- More carbon goes back to the soil in organic farming, which also stimulates productivity. As agricultural models with less pollution are applied on a large scale, the negative impact of agriculture on nature, especially on natural resources, will decrease.

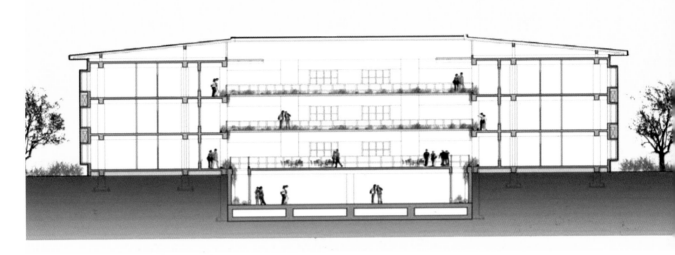

Location: Chongzhou, Sichuan, China.
Established: October 2014.
Area: 620 ha (1,532 ac). Housing area 3.6 ha (9 ac), surrounded by 16 ha (40 ac) of farmland. Density of buildings 40%, green coverage 46%.
Population: Current community members are 84 families, more than 150 people, adults aged 30–50, and some children.
Housing: 150–180 dwellings with diversity between single-family dwellings, terraced town houses, apartments and shared houses.
Communal buildings: Two residential buildings with 84 families. Will be fully developed with seven residential buildings and 294 families.

Following my inner voice

by Yeming He / resident of Hua Tao ecovillage

Like many other members, my Hua Tao story began with the preparation for an international forum. I was working in Peking University and the forum was the last major event of our research program in late 2013. The forum was a big success, and an incubator for several practical initiatives. Hua Tao ecovillage, which is now being realized with fine facilities and enjoyable living conditions, was one of them. At the forum, I truly felt the power of innovating together for the first time. What happened afterwards has shown how a new theoretical system can come into real life from nothing.

In Hua Tao, I took a big step towards my inner growth, which has affected not only my external life, but also deeply affected my whole family. At the beginning, I served as a secretary and held meetings with the founders of Hua Tao ecovillage from different parts of China once a month. Most of the meetings took place on weekends and we all had to pay for our own travel expenses. We flew to meetings and discussed the implementation scheme. Meanwhile sharing sessions were held in certain cities. The start of Hua Tao ecovillage was a very formal and practical process. Life started to be simple as we lived and worked together in harmony. My world also became more pure and beautiful. I gradually found that the power of love would spread rapidly among people and finally flow back to me. Hua Tao means more than the heritage of Tao in Chinese civilization; we follow the Tao to fulfill our commitment of caring about public welfare through spiritual practice. That is what I learned through the process of building Hua Tao ecovillage."

Karise Permatopia Denmark
A holistic and visionary project

by Mikkel Klinge

Karise Permatopia is making a mutually beneficial connection between housing, farming and the production of renewable energy, with closed resource cycles, to make a self-sufficient, environmentally sustainable community with low living costs.

Karise Permatopia will have the best of private ownership, cohousing, rental housing, communes, ecovillages, community supported agriculture, professional ecological farming, permaculture and business networks, that are connected in a holistic and well-planned visionary project, for the mutual benefit of its inhabitants and the natural systems.

The old farm is the natural meeting place for Karise Permatopia. The building is being renovated as a common house, laundry and farm office. In the long term, it will also have office space, meeting rooms, guest rooms, play areas, a café, farm shop and workshops. There will be a car share system, with electric cars charged with cheap electricity from the onsite windmill.

Housing ownership can vary between cohousing, self-ownership and renting, on commonly owned land. The village is organized in eight housing clusters with 90 houses. There are five housing types, ranging from small two-room flats to the largest six-room apartments. The common house, farm, infrastructure and car share system will be organized as four separate entities, where all households own a share.

Karise Permatopia is being developed by experts in close dialogue with future inhabitants and households on a waiting list. Participation creates a feeling of ownership and social connection before people move in. Since mid-2014 there have been several workshops for the members' association. There are groups of members working on housing, the central heating system, the common house, the many aspects of the farm, social organization, learning and teaching, decision-making processes, welcoming new members, internal communication and coordination. In 2017, their focus is on renovating the common house, trying out various decision-making processes, and finalising the social organization of the project.

House Arkitekter from Copenhagen, with project leader Søren Olsen, are the project's architects, and responsible for coordinating the layout design and housing.

Housing

The modern ecological terrace houses have a combination of plentiful daylight, good spaces both indoors and outdoors, a healthy indoor climate, environmentally friendly materials, low energy use, low maintenance, long life, and because of a rational planning and building process, low prices.

All the houses have the same structure and materials. The basic module is a two-storey space with a mezzanine floor that has a bedroom/living room facing a common courtyard, with an open plan to the kitchen/living room. Under the mezzanine, there is an entrance with a bathroom. The houses are made of wood with prefabricated bathrooms, wall elements, floors and roof elements. Houses have a moisture diffusive outer wall and roof, and nontoxic healthy building materials. The interior is plasterboard painted with natural paints, oiled birch parquet floors and sound insulation with wood cement panels. Outer walls are prefabricated elements insulated with wood wool mats and paper granulate, and clad with spruce panelling treated with linseed oil. To save money there are standard prefabricated doors and windows.

The houses are low energy standard, with plenty of daylight, that comply with 2020 Danish building standards. All rooms have floor heating and mechanical ventilation with a heat exchanger. Windows are three-layer energy glass, with 395 and 495 mm insulation in walls and roofs. Roofs have roofing paper that allows collection and use of rainwater.

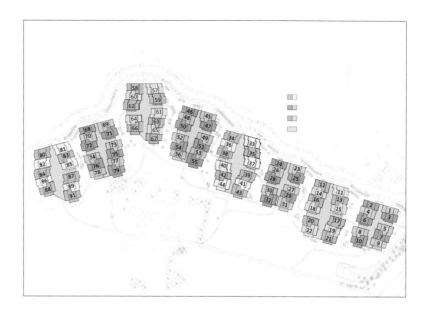

Infrastructure

The technical infrastructure is based on sustainable, well-tested and low maintenance technology, and will provide all necessary services for two-thirds of the normal price, apart from imported drinking water from the local waterworks. Rainwater will be collected from the roofs, filtered and used for clothes washing, toilets and irrigation.

All houses have separation toilets that reduce water use, and wastewater goes to a closed system using willows. Urine from the separation toilets, composted willow, and organic kitchen waste are manure for the farm. This means that the farm will not need any additional animal manure or other forms of fertilizer. Waste separation, composting and recycling will reduce the need for renovation.

Karise Permatopia has a central heating system for space heating and hot water with 9,000 m of earth pipes, and heat pumps run on electricity from a 47m high

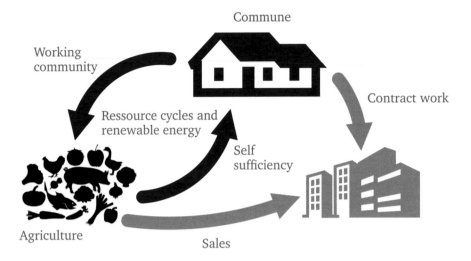

Working community · **Commune** · **Contract work** · **Ressource cycles and renewable energy** · **Self sufficiency** · **Agriculture** · **Sales**

wind generator. A central hot water storage has enough capacity to deliver space heating also in periods with no wind. Electricity comes from the same onsite wind generator, and the need for electricity is reduced with energy efficient lighting, equipment and household appliances. Common facilities, for example, communal meals and laundry, will further reduce energy use. There is a rapid internet connection, IP phones, and TV via fiberoptics.

Community-supported permaculture farm

The farm will be run by ecological farmers well versed in permaculture, with a few hours work a week from the residents. This will allow people to be self-sufficient in sustainable food for about half the normal price.

There will be a large variety of fruits and vegetables produced on about 8 ha (20 ac), with 1,500 m² (16,150 ft²) of greenhouses and a heated greenhouse

(bio shelter) for seedlings. After about three years, there should be a surplus of fruit and vegetables that can either be exchanged or sold. The greenhouse has waste heat from the central heating system and compost, and uses the CO_2 from the composting process to increase plant growth. The farming process will be free of fossil fuels, and run on electric machines that allow for an intensive, high yield production.

Orchards will provide almost all the needed fruits, nuts and berries. A flock of 225 hens will give a surplus of eggs that can be sold in the farm shop, and 500 chickens and a few pigs will cover some of the need for meat. The animals will be an integral part of the farm management, with mobile hen tractors that help reduce pests, provide manure and till the soil.

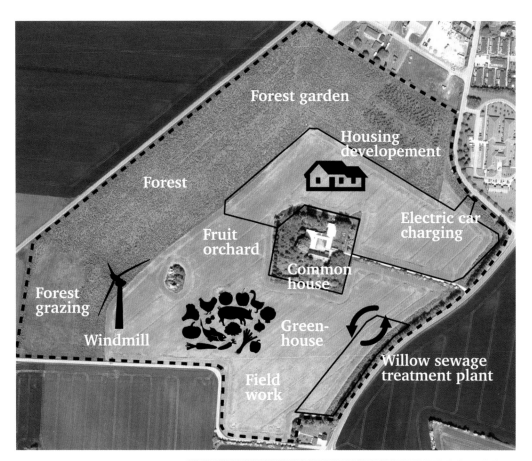

Forest garden

Housing developement

Forest

Electric car charging

Fruit orchard

Common house

Forest grazing

Windmill

Green-house

Willow sewage treatment plant

Field work

Location: Karise south, one hour south of Copenhagen. Walking distance to Karise Station and town with local services.

Established: In autumn 2017, the first people move in and the common house will be finished. Farming starts in 2018.

Area: 4.7 ha (11.6 ac) for housing, 24.3 ha (60 ac) farmland and small ponds, with 12 ha (30 ac) forest, 10.8 ha (26.7 ac) arable land, and 1.5 ha (3.7 ac) communal farm, and gardens.

Population: About 140 adults and 60 children when fully developed.

Housing: 90 houses. Mixed ownership. 90 terrace houses in eight clusters, of which 44 are for rent, 23 cohousing and 23 self-ownership. Ca. 40% of the housing is for families, 30% middle aged, 20% young people and 10% retired.

Communal buildings: The old farmhouse will be a common house, laundry and farm office. In the long term also office space, meeting rooms, guest rooms, play area, café, farm shop and workshops. Car share system.

Contributors' Biographies

Guðmundur Ármann Pétursson, **Solheimar**, born in 1969. He is a graduate in Business Administration from Bifröst University, in Biodynamic Agriculture from Emerson College, and has an MSc in Architecture, Energy & Environmental Studies from the University of East London—Centre for Alternative Technology. He is a member of the Council in Grímsnes-og Grafningshreppur Municipality. He has been the director of Sólheimar since 2005, but with various other tasks in Solheimar since 1988.

Graham Meltzer, Findhorn, has been a passionate advocate of communal living for over 50 years and lived for extended periods in several (including 3 years on a kibbutz, 8 years in an Australian hippy commune and, most recently, 11 years at Findhorn). His doctoral research focused on the environmental advantages of cohousing and he's authored three books on intentional communities of different kinds. Graham currently works for the Findhorn Foundation as both a conference organizer and architectural designer.

MARTI is a writer, photographer, and social-environmental activist, who came to **Auroville** more than 25 years ago. Part of GEN since its formal inauguration, as a GEN Wisdom Keeper, chair of the GEN International Advisory Council, a GEN Elder and GEN delegate to the United Nations. She is co-founder of 'Children and Trees', creating environmental education materials; initiated 'Greenland Spirit', uniting Greenland artists concerned with identity and climate change; and is the author of *This Earth of Ours*.

Macaco Tamerice (Martina Grosse Burlage), has lived in **Damanhur** since 1993, and is the coordinator of International Community Relations, and President of the non-profit Association Damanhur Education. She has been Vice President and President of GEN-Europe, and is on the Advisory Board of GEN-International. Macaco is an international speaker, facilitator and Gaia Education Educator. Fluent in five languages, trained in community building and conflict resolution during the last 23 years, Macaco has held many roles of social responsibility in Damanhur.

• **Christine Arlt** works in **SEKEM**'s Public Relations Department, with SEKEM's main activity fields (economy, ecology, cultural and social life). SEKEM is a wonderful and unique place to work in and of course to visit!
• **Christina Büns** recently joined SEKEM's Public Relations Department, and finds it inspiring to work and live with people who are passionate and dedicated to fostering SEKEM's meaningful mission—day by day.

• **Andreas Kamp**, 38, has lived with his family at **Svanholm** since 2013, working both in Svanholm's administration and as an environmental sustainability assessment researcher. He is involved in a strategic approach to sustainable development at Svanholm based on The Natural Step framework, and the transition to a non-fossil fuel-based society.

• **Mira Illeris** and **Esben Schultz** moved to Svanholm in 2011, to establish a cluster of permaculture farms on Svanholm land. Qualified as organic farmers and permaculture designers, they also work as permaculture teachers, and publish the Nordic Permaculture Magazine.

 **Giovanni Ciarlo** is a sustainability consultant, musician and change agent. He served on the faculty of the Sustainable Businesses and Communities MA program at Goddard College, and presently works in the management of Gaia Education Design for Sustainability (GEDS). Past-president of the Board of GEN, and cofounder of the **Ecoaldea Huehuecoyotl**. His work emphasizes team building, deep democracy, social justice and environmental responsibility.

• **Ana Carolina Beer Simas** works in sustainable education and collaborative communication, lecturing at the Federal University of Viçosa, Brazil. She works with **Céu do Mapiá** and communities in the Purus National Forest.

• **Felipe Nogueira Bello Simas** has a background in agronomic engineering and doctorate in soil sciences. He teaches natural sciences at Universidade Federal de Viçosa, Brazil.

They held the AmaGaia—Ecovillage Design Education in the Purus National Forest. They live in a small ecovillage project in Minas Gerais.

Alex Cicelsky is a founding member of **Kibbutz Lotan** and the Center for Creative Ecology (CfCE). He studied Soil and Water Science at the Hebrew University and Architectural Research in Energy Efficient Housing in Hot Climates at Ben Gurion University. He has designed many of the buildings and ecologically sound technologies in use in the CfCE as well as consulting in green design throughout Israel. Alex lectures worldwide on how ecovillages and environmental initiatives play a part in conflict resolution.

Lucilla Borio is a founding member and resident of **Ecovillage Torri Superiore**. In 1999/2004 she was the GEN-Europe secretary and CEO (Middle East and Africa), GEN-International Chairperson and UN delegate. She works with meeting facilitation, group dynamics and decision-making methods, offering support and consultation to consensus-oriented groups. She is the IIFAC (International Institute for Facilitation and Change) representative in Italy, and trainer of the CLIPS programme (Community Learning Incubator Programme for Sustainability). She speaks Italian, English, French and Russian.

Liz Walker lives with her husband at **EcoVillage Ithaca**, which she co-founded with a colleague in 1991. She has worked as executive director of its non-profit arm, Learn@EcoVillage, and as development manager for each of the three cohousing neighborhoods. She was on the founding board of Gaia Education and has written two books, EcoVillage at Ithaca: Pioneering a Sustainable Culture, (2005, New Society Publishers) and Choosing a Sustainable Future: Ideas and Inspiration from Ithaca, New York, (October, 2010, New Society)

Ave Oit is the long-time chairwoman and the engine behind Lilleoru NGO, one of the original initiators of the Network of Estonian Eco-communities, and together with Ingvar Villido was one of the founders of **Lilleoru**. Conscious human development and sustainable lifestyle have been central to her activities ever since. She helped to establish the eco-store chain Biomarket in Estonia and is now a training manager there. Ave loves North American native culture and spirituality.

Leila Dregger, freelance journalist and teacher of peace journalism, was born in 1959. She has written the books *Tamera: a Model for the Future, Frau-Sein allein genügt nicht—mein Weg als Aktivistin für Frieden und Liebe (Being a Woman is not Enough—my Path as Activist for Peace and Love), Desert or Paradise* (authored with Sepp Holzer), and *Ecovillage: 1001 Ways to Heal the Planet* (with Kosha Joubert). She lives and works mainly in Tamera.

Martin Stengel, 50, is an appropriate technology engineer and a permaculture designer. He has co-designed and co-managed the ecovillage **Sieben Linden**, where he lives with his family (two children) in a self-built straw-bale house. He teaches Ecovillage and Permaculture Design and courses on community building. He has coached intentional communities and trains human resources managers in communication and coaching. He currently supports a British company in design for regenerative supply and resilient livelihoods in Asia and Africa.

John Talbott has been the Project Director for the **Narara Ecovillage** since its founding in 2012. He was the Director of the Findhorn Ecovillage Project for more than 20 years, where he helped pioneer green building methods, ecological infrastructure and renewable energy systems. He is the author of *Simply Build Green* and lectures widely on the subject of building sustainable communities.

Simen Torp, 39, lives in **Hurdal Ecovillage** with his three children. As third generation eco-pioneers, growing up on an organic farm in Oslo, he learnt from the 'school of life'. At age 16, Simen had a strong vision to create Norway's first new ecovillage, Hurdal, and numerous ecological companies like Aktivhus (natural house company), Filago (ecovillage company), Green Sleep Norway (organic mattresses), Bioking Norway (organic food), Kjøkkenhagen (cafe, bakery and health food), Feel Real and Win-Win.

• **Fish Yu** is a volunteer, and facilitator of Pachamama. She likes to live a simple life, with genuineness, kindness and beauty.
• **Fanzhi** is a volunteer, and translator in Beijing Sustainable Development Committee. Nature is as wild as it is peaceful, and so is a human heart.
• **Woody Cui** is a member of **Hua Tao**. Awakened by nature, he is going through a transformation process from being a professional timid ego to a happy co-creator with Mother Earth.

Michael Farrelly currently serves as Programme Manager at Tanzania Organic Agriculture Movement. With a masters degree in rural development and 14 years experience of development work across Africa, from building village water supplies to promoting ecological agriculture, he has designed several climate change adaptation projects. The acclaimed **Chololo** ecovillage project brought together seven diverse organizations to create a model of good practice in climate change adaptation in dryland Africa.

The **Tasman Ecovillage** chapter is a group effort, coordinated by **Karen Weldrick** and **Jane Morton**.
• Karen has lived at Tasman Ecovillage since its inception, and is a founding member of the residents' association. Fascinated by life in co-operative culture, she is passionate about preserving and restoring natural values.
• Jane is a clinical psychologist campaigning for action on the climate emergency. She owns an apartment in the village and hopes to retire there after doing her bit to preserve a liveable planet.

Mikkel Klinge was born in 1973, studied economics and philosophy, and has worked with sustainable transition. He has developed a monetary and tax reform which reduces the inherent growth imperative and inequality of our contemporary economic system, and makes sustainable transition of businesses possible on a national level. He has also developed an exchange and interest-free banking model. In the autumn of 2012 he got the idea for **Karise Permatopia**, and has since then been the main developer of the project.

Frederica Miller, editor
To edit the book Hildur chose the Norwegian/English eco-architect Frederica Miller, Gaia-Oslo. Since the 1980s she has worked with the ecologically sound design of housing, and other projects such as urban sustainable villages and ecovillages. She is a qualified architect and permaculture designer who also works with social participation processes.

GAIA EDUCATION
Design for sustainability

Brief history

Gaia Education was officially launched in 2005, after seven years of preparatory work by some of the world's leading ecovillage pioneers, permaculture designers and sustainability educators and practitioners. Since then our work has expanded to reach nearly 14,000 graduates from 104 different countries through our collaboration with over 80 partner organizations in 49 countries on six continents. After successfully contributing to the United Nations 'Decade of Education for Sustainable Development', we were invited to join the UNESCO 'Global Action Plan' in 2015 and are actively supporting the implementation of the Sustainable Development Goals (SDGs) at community level around the world.

Gaia Education mission & vision

Gaia Education is a leading-edge provider of sustainability education that promotes thriving communities within planetary boundaries.

We work towards a resilient future within planetary boundaries where no one is left behind. A world of safe and nutritious food; of clean drinking water; of universal access to sustainability education; of physical, mental and social well-being. A world that uses energy and materials with greater efficiency, distributes wealth fairly and strives to eliminate the concept of waste. A world of universal respect for human rights and human dignity; of justice and equality; of respect for race and ethnicity; and of equal opportunity permitting the full realization of human potential while promoting shared prosperity.

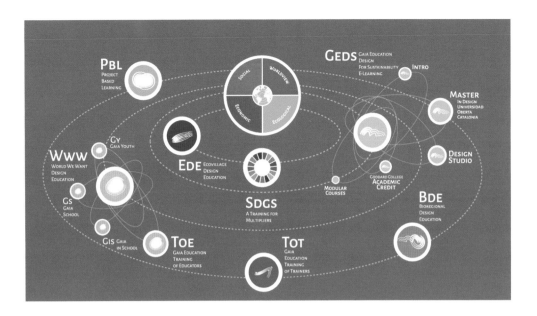

Gaia Education learning universe

The first course offered by Gaia Education and still the flagship of our face-to-face programmes is the 125-hour 'Ecovillage Design Education' (EDE) training in participatory whole systems design for sustainable communities and transition initiatives. It offers an excellent starting point to your learning journey with Gaia Education. Our online courses include a short introduction to the 'Big Picture' and the in-depth Design for Sustainability (GEDS) programme. Our

'Project Based Learning' (PBL) courses focus on practical capacity and resilience building for communities under severe threat of climate change impacts, as well as on supporting migrants, refugees and unemployed youth. Our Training of Trainers (ToT) support those who want to become certified trainers, facilitate transformative learning environments and host multi-stakeholder conversations that enable collaborative action. Courses on community scale implementation of the SDGs and a blended learning course in 'Bioregional Design Education' (BDE) are the most recent programmes adding to the Gaia Education 'universe' of programmes.

4-D Framework for integrative whole systems design

At the heart of Gaia Education's diverse range of courses lies the 4-D Framework for integrative whole systems design. It transcends and includes the conventional three dimensional model of sustainability aiming to integrate social, ecological and economic concerns in the creation of sustainable solutions, by taking a more holistic approach that also addresses the critical importance of worldview and value systems.

SDGs

Gaia Education, in partnership with UNESCO Global Action Programme, has been conducting 'Achieving the Global Goals—One Community at a Time, a Training for Multipliers'. This highly interactive and participatory workshop invites participants to engage in progressive conversations about the local relevance of the SDGs based on a set of SDGs Community Implementation Flashcards containing more than 200 questions structured into the four dimensions of Gaia Education's whole systems approach to sustainability.

E-Learning Programmes

Gaia Education Design for Sustainability (GEDS) e-learning course is an excellent way for professionals, educators, academics, activists and community organizers to deepen in their holistic understanding of integrative whole systems design. It offers a wide range of inspiring case studies, methodologies and practical tools to promote regenerative development and facilitate sustainable community design at the local and regional scale.The intensive programme requires a minimum of 400 study hours and is structured into four dimensions that can be taken separately—social design, ecological design, economic design, and worldview—followed by an additional design studio in which students revise what they have learned by applying it to a real world design challenge within collaborative design teams. The course is offered in English, Portuguese and Spanish—the latter with the option of working towards a Masters degree at the Universidad Oberta de

Catalunya (UOC).

CASE STUDY 1: BANGLADESH
Name: Building Capacity & Empowering
Communities in Khulna & Bagerhat
Districts, Southern Bangladesh, towards
Sustainable Agriculture, Aquaculture
Development and Climate Change
Adaptation Interventions
Period: 4 Years
Funded: Scottish Government

CASE STUDY 2: SENEGAL
Name: Increasing Food Security,
Income Generation and Environmental
Sustainability in the Podor Region,
Northern Senegal
Period: 3 Years
Funded: UK AID

CASE STUDY 3: INDIA
Name: Empowering and Building
Capacity of Tribal Communities to
Increase Food Security, Social Cohesion
and Climate Resilience
Period: 2 Years
Funded: Scottish Government

CASE STUDY 4: ITALY
Name: Socio-Economic Integration of
Migrants and Unemployed Youth through
Organic Products
Period: ongoing
Funded: LUSH, Global Whole Being Fund
of RSF Social Finance. Eileen Fisher,
Voelkel. Permakultur Akademie

GLOBAL ECOVILLAGE NETWORK
Fueling the power of community for a regenerative future

We are called to re-own our capacity to be guardians and healers of life and thus truly come home to the planet and ourselves. In GEN, we believe that every community has the right to protect and the capacity to restore their environments. We believe that we can heal our past and invite a future that is in alignment with natural law. We believe that a world that lives within its means and builds bridges of solidarity across all borders is not only possible, but true to our very nature.

Today, GEN reaches out to around 10,000 communities on all continents and inspires governments around the world to include ecovillages in their strategies for the implementation of the SDGs and Climate Agreements.

We have learned that, while there is no one way of being an ecovillage, there are three core practices shared by all:

- Being rooted in local participatory processes
- Integrating social, cultural, economic and ecological dimensions in a whole systems approach to sustainability
- Actively restoring and regenerating their social and natural environments

The following infographics illustrate some of GEN's impact in the world. For more information please visit **www.ecovillage.org.**

30
consultancy
experts

300
sustainability
solutions

23
governments
interested in GEN

78
ambassadors

2821
volunteer hours

26
partnership organizations

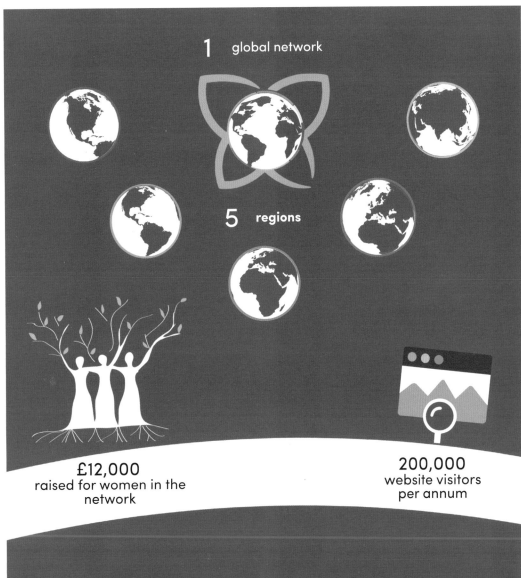

1 global network

5 regions

£12,000
raised for women in the network

200,000
website visitors per annum

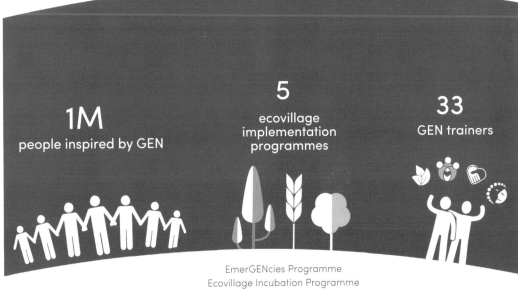

1M
people inspired by GEN

5
ecovillage implementation programmes

33
GEN trainers

EmerGENcies Programme
Ecovillage Incubation Programme
Urban Eco-Neighbourhoods Programme
Ecovillage Transition/Development Programme
Greening Schools for Sustainable Communities Programme

The Sustainability Mandala
A holistic approach to community design

The Global Ecovillage Network embraces a holistic approach to sustainability, integrating the social, cultural, ecological and economic dimensions of existence. At the centre, we place the practice of whole systems design.

The dimensions of sustainability and the central path of whole systems design make up the Sustainability Mandala—our road map to the creation of ecovillages—using participatory processes to integrate ecological, economic, social, and cultural dimensions of sustainability in order to regenerate social and natural environments.

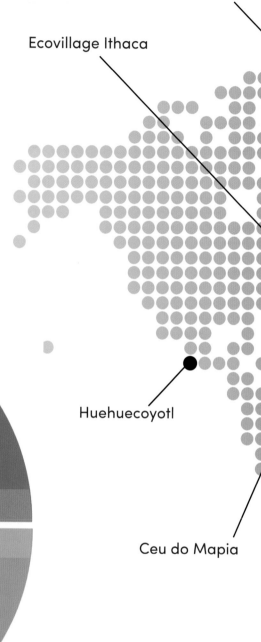

Solheimar

Ecovillage Ithaca

Huehuecoyotl

Ceu do Mapia

SOCIAL

ECONOMY

CULTURE

ECOLOGY

WHOLE SYSTEM DESIGN

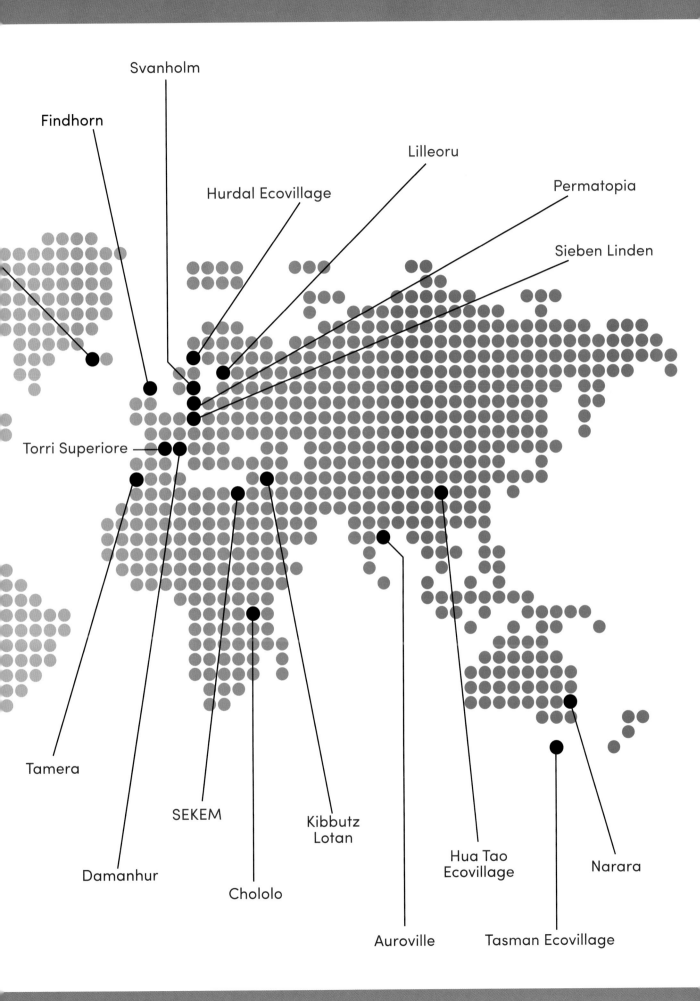

Findhorn

Svanholm

Hurdal Ecovillage

Lilleoru

Permatopia

Sieben Linden

Torri Superiore

Tamera

Damanhur

SEKEM

Chololo

Kibbutz
Lotan

Auroville

Hua Tao
Ecovillage

Tasman Ecovillage

Narara

United Nations Sustainable Development Goals

by May East

We are faced with the challenge of collectively re-designing the human presence on Earth. Now is the time for transforming humanity's planetary impact from predominantly degenerative to by and large regenerative! It is our generations, those alive today, who face the task of regenerating the healthy, life-supporting functions of marine and terrestrial ecosystems everywhere. In doing so we will create the basis for thriving local communities and vibrant circular economies. We can create a fairer distribution of resources through widespread global-local collaboration while learning to live within planetary boundaries. This is the promise ahead, if we come together across sectorial, national and ideological divides to collaborate in implementing the United Nations Sustainable Development Goals at the local, regional and global scale. It is time to get to work – one community at a time!

Four Keys to the Design of Sustainable Communities

This book is part of the Gaia Education Four Keys to the Design of Sustainable Communities series, offering an overview of the state of the art of sustainable settlements globally. Each key covers one of the dimensions of the Ecovillage Design Education curriculum: Economic, Worldview, Ecological and Social.

Other titles in the series:

Social Key
Beyond You and Me: Inspirations and Wisdom for Building Community
Editors: Kosha Anja Joubert, Robin Alfred
A practical people care anthology for anyone seeking to build community and embrace diversity.

Ecologic Key
Designing Ecological Habitats: Creating a Sense of Place
Editors: E. Christopher Mare, Max Lindegger
An eloquent exploration of regenerative approaches addressing limits to growth, climate change and resource depletion.

Economic Key
Gaian Economics: Living Well within Planetary Limits
Editors: Ross Jackson, Helena Norberg-Hodge, Jonathan Dawson
Explores how we can develop healthy and abundant societies in harmony with our finite planetary resources.

Worldview Key
The Song of the Earth: A Synthesis of the Scientific and Spiritual Worldviews
Editors: Maddy Harland, Will Keepin
Brings together the voices of leading visionaries in science, spirituality, indigenous wisdom and social activism.

References

Solheimar, Iceland: http://www.solheimar.is/en/, Picture credits: Pétur Thomson

Findhorn, Scotland: www.findhorn.org, https://www.facebook.com/findhornfoundation, https://twitter.com/FindhornFound. Picture credits: Graham Meltzer, John Talbott, Findhorn Foundation
Contact information: Graham Meltzer, graham.meltzer@findhorn.org

Auroville, India: http://www.auroville.org/. Picture credits: All photos by MARTI
Contact information: MARTI, marti@auroville.org.in

Damanhur, Valchiusella Italy: http://www.damanhur.org/nb.
Contact information: Macaco, macaco@damanhur.it
All photos from Damanhur's archive.

SEKEM, Egypt: http://www.sekem.com/en/index/. Picture credits: ©SEKEM

Svanholm, Denmark: http://svanholm.dk/.
Picture credits: Andreas Kamp, Josephine Høgenhaug Aaen, Esben Schultz and Mira Illeris.

Huehuecoyotl, Mexico: http://huehuecoyotl.net/. Picture credits: Svante Vanbart and Giovanni Ciarlo
Contact information: Giovanni Ciarlo, giovanni@ecovillage.org

Ceu do Mapia, Brazil: http://www.santodaime.org/site/site-antigo/comunidade/visit01.htm
Picture credits: Rafael Oliveira, AmaGaia (circle)

Kibbutz Lotan, Israel: www.KibbutzLotan.com, Facebook—Kibbutz Lotan and Center for Creative Ecology
Picture credits: Center for Creative Ecology, Alex Cicelsky, Mark Naveh, Gwen Scully, Dale Lazar, Ido Zevulun

Torri Superiore, Italy: http://www.torri-superiore.org/en/
Picture credits: Drawing by Simona Ugolotti, 2015, Genova, "Greetings with no border". Ecovillage plan by Studio Architetto Gianfranco Fava, 1993, Turin. "Aerial view of the Torri village and Ecovillage" Torri Superiore archives. "Frontal View of the Ecovillage in 2013" Nina Freund, Ecovillage Torri Superiore archives. "Olive harvest and the ancient olive trees" Nina Freund, Ecovillage Torri Superiore archives. "Celebration for the 25th anniversary of the Ecovillage, 2014" and "The Bevera River and the swimming holes" Lucilla Borio, Ecovillage Torri Superiore archives

EcoVillage Ithaca, USA: http://ecovillageithaca.org/
Picture credits: TREE Common House: Jim Grant. FROG Common House: Jeff Gilmore. FROG—drawing by Jerold Weisburd, architect. TREE—drawing by Rick Manning, landscape architect. Lake cover picture and all other photos: Jim Bosjolie

Lilleoru, Estonia. : www.lilleoru.ee, https://www.facebook.com/lilleoru/
Picture credits: Flower of life park, Yantra photo and community photo by Mari Kadanik
Winter photo and Temple photo—Aimar Säärits

Tamera, Portugal: https://www.tamera.org/index.htmlwww.tamera.org
"Tamera—a model for the future", by Leila Dregger, and more books from Tamera at www.verlag-meiga.org
Picture credits: Welcome ceremony around Lake 1. Women power—Photos: Nigel Dickinson
The main valley of Tamera before Lake 1—and after. Photos: Thomas Lüdert and Simon du Vinage
Auditorium of Tamera, community of Tamera, Cultural Center, Logo of Tamera, Visitors at the shores of Lake 1, Escola da Esperanca, the school, The love school. Photos: Simon du Vinage. Experimental building. Photo: Tamera Archive. Different views in the Solar Testfield and Solar Kitchen. Photographers: Leila Dregger Bernd Eidenmüller, all others: Simon du Vinage

Sieben Linden, Germany: http://www.siebenlinden.de/
Picture credits: Sieben Linden residents and visitors
Contact information: martin.st@siebenlinden.de

Chololo, Tanzania: https://chololoecovillage.wordpress.com/
Picture credits: Shoes and goats by Kerry Farrelly. Woman in field by Dr Francis Njau. All other photos by Michael Farrelly
Contact information: Michael Farrelly, mrfarrelly@gmail.com

Narara, Australia: http://nararaecovillage.com/
Picture credits: The 'nature' shots are by Richard Cassels. Otherwise John Talbott.
Contact information: John Talbott, info@nararaecovillage.com

Hurdal, Norway: https://www.hurdalecovillage.no/
Picture credits: Plans and illustrations Aktiv-hus architects and Rolf Jacobsen Gaia Architects.
Photos: Filago.
Sustainable Urban Village © Helen and Hard architects

Tasman Eco Village, Australia: http://tasmanecovillage.org.au/
Picture credits: ©Nilmini De Silva 2015 and Neil Robertson
Contact information: info@tasmanecovillage.org.au

Hua Tao, China: http://www.huataoecovillage.com/en/index.html
Picture credits: Ms. Wang Zeng and Mr. Liu Fei.

Karise Permatopia, Denmark: http://www.permatopia.dk

Picture credits: First ground breaking—photo Bjørn Hernes.
Drawings by House Arkitekter Copenhagen Architect Søren Olsen

FINDHORN PRESS

Life-Changing Books

Learn more about us and our books at
www.findhornpress.com

For information on the Findhorn Foundation:
www.findhorn.org